THE CHEMISTRY

OF

ELECTRODE PROCESSES

THE CHEMISTRY

OF

ELECTRODE PROCESSES

ILANA FRIED

Department of Inorganic and Analytical Chemistry
The Hebrew University of Jerusalem, Jerusalem, Israel

1973

ACADEMIC PRESS LONDON AND NEW YORK
A subsidiary of Harcourt Brace Jovanovich, Publishers

CHEMISTRY

ACADEMIC PRESS INC. (LONDON) LTD.
24/28 Oval Road,
London NW1

United States Edition published by
ACADEMIC PRESS INC.
111 Fifth Avenue
New York, New York 10003

Library of Congress Catalog Card Number 73–9458
ISBN: 0 12 267650 5

Printed in Great Britain by
C. F. Hodgson & Son Ltd., 50 Holloway Road, N7 8JL, England

Preface

The name " electrodics " was first used by Professor Conway and has since been adopted by many. It is that part of science which deals with the transfer of electric charge between a solid and a liquid phase and it is the subject of this monograph.

The purpose of this book is to present the field of electrodics to the advanced student or the qualified chemist. It is assumed that the reader is conversant with thermodynamics, basic kinetics and the basic concepts of conductance. No familiarity with electrochemistry is assumed. The book starts with very basic concepts of electrochemical cells, of potential and of kinetics, and develops these according to the needs of the study of electrode processes.

My purpose throughout was to present the subject-matter simply rather than rigorously. Therefore, the equations were sometimes stated rather than derived; other equations were arrived at by analogy rather than directly. The literature on electrodics has grown rapidly in the last few years: reviews and books on every aspect of the field have been published. This book is an introduction to electrodics and is meant to enable the reader to use the more advanced literature intelligently. The Bibliography includes a selection of books that I found useful; it ranges from very simple ones to very sophisticated, but is by no means complete.

The list of symbols includes only the basic ones; these have been used with several subscripts or superscripts in several ways. The exact meaning of the symbol used is given in the text.

The International System of Units (SI units) has been used. Its merits are unquestionable: it removes the use of conversion factors, especially in energy measurement; it does away with the use of mixed units, such as amperes per cm^2 (amperes belongs to the mks system, and cm to the cgs system); it does not use arbitrary units such as the calorie. The disadvantage of using this system is that the familiar numbers, such as bond energies, current densities etc., look different. While this difficulty is real, it can be readily overcome by a few comparison exercises.

I want to thank Dr. W. D. Cooper of the University of Edinburgh for giving this book many hours of his time. I also want to thank Dr. R. Parsons of the University of Bristol for reading the manuscript and for many enlightening remarks. Last I want to thank my family for their wonderful encouragement and patience.

ILANA FRIED

Jerusalem, September 1973.

v

Acknowledgements

Permission to reproduce the following Figures is gratefully acknowledged: Figure 20: J. O'M. Bockris, M. A. V. Devanathan and K. Muller, *Proc. Royal Soc. (London)*, **A274** (1963), 55 (figure 8). Copyright by the Royal Society. Figure 26: D. C. Graham, *Chem. Rev.*, **41** (1947), 441 (figures 5, 6). Copyright by the American Chemical Society. Figures 28, 29: D. C. Graham, *J. Am. Chem. Soc.*, **76** (1954), 4819 (figures 1, 2, 3, 4, 5). Copyright by the American Chemical Society. Figure 30: D. C. Graham, *J. Am. Chem. Soc.*, **79** (1957), 2093 (figure 6). Copyright by the American Chemical Society. Figure 32: J. O'M. Bockris and A. K. N. Reddy, *Modern Electrochemistry*, vol. 2, Plenum Press, New York, p. 709. Copyright by Plenum Publishing Corporation. Figure 34: D. C. Graham, *J. Am. Chem. Soc.*, **80** (1958), 4201 (figure 1) and *ibid.*, **83** (1961), 1291 (figure 4). Copyright by the American Chemical Society. Figure 35: D. C. Graham, *J. Am. Chem. Soc.*, **83** (1961), 1291 (figure 3). Copyright by the American Chemical Society. Figure 36: D. C. Graham, *J. Chem. Phys.*, **22** (1954), 449 (figure 11). Copyright by the American Institute of Physics. Figure 39: A. N. Frumkin, *Electrochim. Acta*, **9** (1964), 465 (figure 5). Copyright by Microfilms International Marketing Corporation. Figure 59: P. J. Elving and D. L. Smith, *Anal. Chem.*, **32** (1960), 1849 (figure 3). Copyright by the American Chemical Society.

Contents

List of Important Symbols

A—surface area

a—activity

b—rate of potential change

C—capacity

c—concentration

D—diffusion coefficient

d—distance between capacitor plates

E—electrode potential vs. a reference electrode

$E°$—standard electrode potential vs. a reference electrode

e—the electron

e—the charge of the electron

F—the faraday

F—force

G—Gibbs free energy

g—acceleration due to gravity

I—current

I_l—limiting current

I_p—peak current

i—current density

i_l—limiting current density

i_p—peak current density

J—flux

K—conductivity

K—dielectric constant

K—equilibrium constant

$\vec{k}$—rate constant of forward reaction

$\overleftarrow{k}$—rate constant of reverse reaction

L—surface concentration

M—molecular weight

M—number of moles

M—weight of a drop

m—rate of flow of mercury

n—number of electrons per unit reaction

n_i—number of particles i in unit volume

n—number of particles in moles

Ox—the oxidised species

P—pressure

Q—amount of electricity passed in a reaction

q—quantity of electric charge

q_m—surface charge on the electrode

q—heat supplied

R—gas constant

R—resistance

Red, R—the reduced species

r—radius of curvature of the meniscus

S—entropy

T—absolute temperature

t—time

t—transference number

U—internal energy

u—mobility

V—voltage

V—volume

v—reaction rate

v—velocity of flow of liquid

W—electronic work function

W—weight

W—Warburg impedance

w—work done by a thermodynamic system

X—mole fraction

Z—impedance

z—ionic charge

$\vec{\alpha}, \vec{\beta}, \vec{\gamma}$—reaction orders for the forward reaction

$\overleftarrow{\alpha}, \overleftarrow{\beta}, \overleftarrow{\gamma}$—reaction orders for the reverse reaction

β—transfer coefficient

Γ—surface excess

δ—thickness of diffusion layer

δ_r—thickness of reaction layer

ε—permittivity

ε_i—energy level of particle i

η—overpotential

θ—fraction of surface covered with adsorbate

θ—contact angle between mercury and capillary wall

θ—resistance of electron transfer

Λ—molar conductance

μ—chemical potential

$\bar{\mu}$—electrochemical potential

v—kinematic viscosity

v_i—order of electrode reaction for substance i

ρ—density of mercury

ρ—volume charge density

σ_m—surface charge density on the electrode

ϕ—the electrode inner potential

χ—the electrode surface potential

ψ—the electrode outer potential

ω—angular velocity of rotation

ω—frequency of alternating current

1. Introduction

A. Definition of the field of interest

Electricity interacts with matter because electrons are part of matter and form the chemical bonds. When electrons are transferred from one molecule to the other we call it a redox reaction. Since electric current is the movement of electrons, "micro" electric currents then exist in the solution where redox reactions take place. If all these micro-currents were made to flow in one direction we should be able to measure them as one "macro" electric current. Batteries (which are also called "galvanic cells" or "voltaic piles") are devices which do exactly this: they produce electric current by making redox reactions take place at electrodes, i.e. at the metal solution boundary. The metal can be either the source or the sink for electrons. Thus electric current is made to flow from the metal into the solution or from the solution into the metal. Can one do the reverse? Can one induce redox reactions by passing through the solution current from a source? The answer is definitely "yes". The instrument by which such changes are produced is an "electrolytic cell". A simple cell can be constructed from two pieces of dissimilar metals dipping into a solution of some electrolyte in a beaker. The metal pieces are now the electrodes. This book is concerned with chemical reactions produced by electric current or electric current produced by chemical reactions at electrodes. It is concerned with redox reactions in cells.

The study of electrode reactions can conveniently be called "electrodics", a name which will be used throughout this book.

An electrode reaction takes place between the electrode and the solution, (which may contain some neutral chemicals and electrolytes). Since the electrode surface is the only place where the solution and the electrode meet, it must be a surface reaction. Therefore it is very important to be familiar with the structure of the interphase, i.e. the region which includes the surface of the electrode and that part of the solution which is influenced by it. Unfortunately, our knowledge of the surface is far from satisfactory. The classical approach will be presented in this book.

So far we have considered cells which produce chemical changes (material) by consuming electric current (energy) or those which produce electric current (energy) by consuming chemicals (material). There is, however, another class of electrodic processes called corrosion which includes those processes consuming both material and energy and are, therefore, very undesirable. Protection of metals against corrosion is one of the applications of electrodics; some of its basic aspects will also be dealt with here.

B. The applications of electrodics

The potential across a cell is very simply related to the change in free energy of the cell reaction

$$\Delta G = -nFE \qquad 1.1$$

Where ΔG is the free energy change, n is the number of electrons in unit reaction, F is the Faraday and E is the cell potential. This close relationship and the direct association between the current flow and the rate of electrode reaction (an explanation of which will be given later), plus the ability to control and measure these two quantities, provide a unique tool for the chemist. To appreciate these unique characteristics, it is advantageous to divide the following outline into basic science and technology.

1. *Basic science*

There are two types of problems which can be helped by electrodics.

The first group includes both organic and inorganic synthetic chemistry. Here the ability to control the energy of the reaction plays a vital part. An example of this is the reduction of 1-bromo-1-nitromethane. This compound can be reduced in three steps, involving three values of potential. The first step, and the least energetic, leads to the corresponding *aci* nitro compound; the second leads to formation of acetaldoxime, and the third step is a further reduction of acetaldoxime. Using conventional reducing agents gives a mixture of two or three reduction products. By adjusting the cell potential at the value most suited to one's needs, one particular compound only may be synthesized.

The field of synthetic inorganic chemistry has only recently begun to use electrolytic reactions and one may look forward to greater sophistication in the making of materials as this field develops.

The second group includes the study of homogeneous and heterogeneous electron-transfer reactions. Electrodic techniques have three distinct advantages: the potential at which the electrode reaction takes place is a measure of the free energy of reaction (hence equilibrium constants), the overpotential (if calculable) can indicate the magnitude of the activation energy and the current at any given potential measures the rate of the reaction. It is also possible to study short-lived intermediate species formed during a reaction, electrolytically.

While it is true that not all homogeneous redox reactions lend themselves to electrodic investigation, it is felt that many of the biochemical processes have similar mechanisms to the corresponding redox electrodic ones. This is supported by the common features of these two kinds of reactions: they take place at room temperature, as opposed to ordinary reactions which often have to be heated or cooled, they do not need conditions of extreme pH and their energy can be easily used. Is it only a speculation that biochemical reactions proceed via an electrochemical mechanism?

2. *Technology*

Electrodics is closely connected with three types of technological processes. The first includes energy consuming, material producing processes, such as the manufacture of chlorine, sodium and aluminium on the one hand, and all electroplating processes on the other; the second includes material consuming, energy producing devices such as batteries and fuel-cells, and the third is the study of corrosion and its prevention.

Electric current has the unique property of being a highly powerful and extremely well controlled device for "pumping" electrons in or out of ions and molecules in solution. Being the best reducing agent it is used in the large scale manufacture of the electropositive metals sodium, aluminium and magnesium, and on a smaller scale, the rest of the alkali and alkaline-earth metals. Being the best oxidizing agent, it is used for the manufacture of chlorine and, to a lesser extent, for the production of fluorine. The ability to control cell voltage and current so that only the wanted processes take place is used in the purification of copper.

Electroplating processes use the article to be plated as an electrode, precipitating the desired coating on this article. The coating is applied for either decorative or anti-corrosion purposes or it may be used to make the article more resistant to frictional wear. The material to be plated, the thickness of the film and the solution to be used are determined by the purpose of the plating.

Since 1800 it has been known that certain combinations of metals and salt solutions produce electric current. Later it became clear that this phenomenon is general and that, in principle, one should be able to make any redox reaction produce electric current (equation 1.1). The more negative ΔG for a reaction, the larger will be the potential of a cell based on it. Many batteries and storage batteries were developed over the years, utilizing those systems which lent themselves most easily to their construction.

The advantages of direct conversion of chemical to electrical energy are twofold. One is the efficiency of conversion and the other is the elimination of air pollution. The common way of obtaining electrical energy is to combust the fuel, obtaining heat. This heat is used to expand a gas which moves a dynamo which produces electricity. Whenever heat is involved in an energy cycle, the efficiency of the whole cycle is given by the efficiency of the step involving heat. This cannot be greater than the Carnot efficiency, i.e. 40%. In addition, a direct process is, in principle more efficient because of its simplicity. A typical efficiency of a battery is approximately 90%.

The second advantage of direct conversion of chemical to electrical energy— the absence of air pollution—make fuel-cells very attractive. While fuel cells do have some wastes, it is reasonable to hope that these wastes can be dealt with more easily than can air pollution. However, available devices use rather expensive materials as sources of energy and one always wanted to use cheap materials, such as gas, oil or coal and atmospheric oxygen for the construction of energy producing devices.

The use of batteries and fuel-cells is increasing and efforts are being made to search for new types of cells that will fulfil the basic requirements of a good energy converter. These requirements are first, that the cost of building and operating should be minimal and second, that the weight and size of the energy converter, together with all auxiliary equipment necessary for its operation, should be made as small as possible. Consider car engines as an example. If engines could be made smaller and lighter, but still retain the same power, cars would be cheaper to produce and run. Larger or heavier engines would increase the cost of keeping a car. Thus the power of an energy producing device for unit weight should be maximum. Experience to date indicates that batteries and fuel-cells operate best when small independent sources of power are necessary, such as for amplifiers on telephone lines, light sources in space craft and the like.

The problem of corrosion is immense. One can say with certainty that all metals, except the noble ones (gold, platinum etc.), corrode to a certain extent. Some of them form oxide layers on the surface which protects them against further attack, but some, notably iron, do not form such a layer. Iron corrosion alone costs every year millions of pounds in protection and replacement. There are two main mechanisms by which corrosion occurs. One is by direct oxidation of the metal by air oxygen. Metal oxides are, as a rule, thermodynamically more stable than the metal and oxygen in their elementary states. This is the mechanism by which metals become covered with oxide even in very dry atmosphere. The second mechanism is an electrochemical one: two electrode reactions take place on the surface of the corroding metal; their products are the undesirable corrosion products. It is easier to understand electrochemical corrosion by considering the corrosion of copper plated iron as an illustration: if the plating is broken, the iron, which is in contact with the copper, is exposed to the atmosphere, to rain and often to traffic fumes. The atmosphere contains carbon dioxide and sometimes sulphur dioxide both of which form electrolytes in rain water. Thus, one has a situation similar to that of a battery: two dissimilar metals dipping into an electrolyte solution. In contrast with an ordinary battery, these metals are short-circuited and current flows from the iron, which is oxidized, to the copper. The result of this process is the corrosion of the iron. Other electrochemical corrosion reactions may take place under different conditions, a discussion of these and of the role of electrodics in understanding corrosion is given in Chapter 6.

We see that electrodics has great potentialities in many fields of chemistry and technology. Many problems, up to now unsolved, could be tackled by electrodic techniques. Let us now see how the field of electrodics developed and try to understand the problems that had to be overcome to appreciate the nature of electrode processes. This might enable one to search through the older literature and make proper use of it.

C. Historical development

That electricity has a striking effect on chemicals was known as early as the middle of the eighteenth century when several investigators found that electricity produced by friction machines could "revivify" (reduce) metals from their oxides. The first electrolysis of water was achieved by Deimann and Paets van Troostwyk in Haarlem in the Netherlands, in 1789. Sparks were passed through a cylindrical vessel containing water and the resulting gases (hydrogen and oxygen) were collected as a mixture at the top of the cylinder.

The history of electrochemistry began, however, in 1800 when Volta described the first voltaic pile. The availability of sources of large currents at small potential enabled electrolyses on a fairly large scale: the first production of the alkali and alkaline earth metals by Davy being a spectacular result. Quantitative measurements of current strengths and quantities of electricity brought the formulation of the laws of Faraday in 1834. The second law, stating that every equivalent of material reacts with the same quantity of electricity formed, together with the studies of Dalton and Gay-Lusac on the weights and volumes of combining substances forms the basis of our present understanding of the atomic nature of matter.

Early research on various types of voltaic piles produced the two important discoveries of the hydrogen-oxygen fuel cell and of the lead-lead sulphate storage battery.

In the latter part of the nineteenth century, the science of thermodynamics was developed and its tremendous importance was realized. It was then that the relationship between cell potential and free energy was derived, thereby affording a simple way of measuring free energy changes. The Nernst equation (1888) which gives the cell equilibrium potential as a function of reactant concentration and a standard potential

$$E = E^0 + \frac{RT}{nF} \ln \frac{[Ox]}{[R]} \qquad 1.2$$

(E is the equilibrium cell potential, E^0 is the standard potential, R is the gas constant, T is the temperature, n the number of electrons in unit reaction, F is the faraday and [Ox] and [R] are the concentrations of the oxidized and reduced species) afforded the determination of standard free energy changes and standard chemical potentials. Here lay a tremendous field of study, where theory matched the available experimental know-how and many electrochemists have studied this field of potentiometry. As we know, the concentration terms in equation 1.2 should be replaced by activities and activity coefficients which give an insight into the structure of solutions. The study of solution electrochemistry (sometimes called "ionics") flourished and the electrodics part was left.

Electrodics were not as exciting at the time as ionics—there was certainly no theory of such universality as thermodynamics, nor were the instruments available in the early part of this century capable of measuring potentials, currents and time intervals to the needed degree of accuracy. The few who were interested in the study of electrode processes could only laboriously measure current-potential curves, establish the experimental relationship between them, wonder why they do not fit thermodynamic calculations and speculate on the nature of the mysterious phenomenon of overpotential.

We know that thermodynamics is a very powerful tool for the study of systems at equilibrium, but electrode processes are systems not at equilibrium; when at equilibrium there is no net flow of current and no net reaction. Therefore electrode reactions should be studied using the concepts and formalities of kinetics. Indeed, the same period that saw the flourishing of solution electrochemistry, also saw the formulation of the fundamental theoretical concepts of electrode kinetics: the work of Tafel on the relationship of current and potential was published in 1905; those of Butler and Volmer and Erdey Gruz, which formulated the basic equation for electrode kinetics, were published in 1924 and 1930 respectively. Frumkin in 1933 showed the correlation between the structure of the double layer and the kinetics of the electrode process. The first quantum mechanical approach to electrode kinetics was published by Gurney in 1931.

However, even though the theory was laid out, experimental work did not follow as soon as expected. This was because studies on electrode processes were made using solid electrodes. Solid, constant surface electrodes are extremely prone to "poisoning" by impurities; if the solutions and containers are not extremely clean, the electrode surface will be covered with adsorbed impurities and the state of the surface could not be reproduced well enough to obtain reproducible and meaningful data.

While research on electrode kinetics met many difficulties, the study of electrode interphase under equilibrium conditions advanced rapidly because thermodynamics could be used with Gibbs adsorption isotherm as the main theoretical tool. The pioneering theoretical studies of the structure of the double layer were done in the last few decades of the nineteenth century and first few decades of this century. Helmholz, Gouy, Chapman and Stern, all published their works before 1930. Here, again, experimental work was slow in progressing for the same reasons that hindered the work on electrode kinetics. It was Graham's work on the double layer using the dropping mercury electrode in the 1940's which stimulated a new interest in precise quantitative measurements. The dropping mercury electrode (see Fig. 12) consists of a column of mercury dropping through a very fine capillary. Thus a fresh, clean surface is being created all the time, so avoiding the problem of electrode contamination.

A great deal of experimental data came from a rather unexpected place. In 1922 Heyrovsky introduced the polarograph. This instrument consists of a cell containing a dropping mercury electrode and a saturated calomel electrode connected to a variable voltage source. The current flowing through the cell is measured with a sensitive galvanometer. The current-potential curves obtained are very useful for qualitative and quantitative analysis in ways which will be explained later. The analytical chemists became interested in polarography and data on electrode reactions on the dropping mercury electrode began to accumulate rapidly. Thus, the theory was all laid out, an experimental tool was available and in time the two were combined. This combination, together with development of instrumentation, resulted in the science of electrodics as it is today, with many technical and scientific applications.

While experimental techniques for the study of both electrode kinetics and the structure of electrode surface are quite well developed, the theory lags far behind. We still do not know the quantitative relation between the parameters of the electrode surface and the parameters of the kinetics on this electrode. We understand very little of either the structure of electrolyte solutions in the concentrations involved, or the structure of the solution in the double layer where a very intense field exists. The detailed structure of the metal surface is also not well known. Most of us still use the classical approach to electrode kinetics, when it is quite obvious that a quantum viewpoint should be adopted.

Workers in the field of electrodics must now focus their attention on the dynamics of the surface, and it is hoped that the nineteen seventies will see great development in these areas.

We have seen that the study of electrode processes (electrodics) is concerned with redox reactions in cells, is related to surface chemistry and reaction kinetics and has very many applications in science and technology. Let us now continue and study the basic principles and the applications of electrodics.

2. The Galvanic Cell, Basic Definitions and Concepts

Every galvanic and electrolytic cell has two electrodes. The one which introduces electrons to the cell is the cathode; the one which removes them from the cell is the anode. In this chapter we will consider some of the basic concepts of cells; the meaning of cell and electrode potentials will be examined in detail. We shall define "electrolytic cells" and "galvanic cells"; discuss the difference between surfaces and bulk of matter; consider the location of the site of the electrode reaction, and the forces and laws which control the flow of current and make one electrode the source and the other the sink for electrons.

A. The cell

A galvanic cell is a device which generates electrical energy directly from chemical reactions. An electrolytic cell is a device by which a chemical reaction is made to take place by using electrical energy. These two devices are the same instrument operated in opposite ways and will be called here a cell. Figure 1 shows two cells connected in series. One operates as a galvanic cell and the other as an electrolytic cell. These are made of suitable containers which are each divided into two halves by a porous membrane. In the cell marked G the two halves contain solutions of copper and zinc sulphate with copper and zinc metal rods dipping into their respective solutions. The electrode reactions are

$$\text{cathode} \quad Cu^{2+} + 2e = Cu \qquad\qquad I$$

$$\text{anode} \quad Zn = Zn^{2+} + 2e \qquad\qquad II$$

If the salt solution is fairly concentrated, say of one molar, a potential difference of approximately 1·1 volts will be produced between the metals. In the cell marked E there are platinum electrodes dipping into solutions which do not contain platinum ions. The right-hand compartment contains ferric and ferrous ions and the electrode reaction is

$$Fe^{3+} + e = Fe^{2+} \qquad\qquad III$$

The platinum acts here merely as an electron membrane, its potential being determined by the concentration ratio of ferric and ferrous ions in the

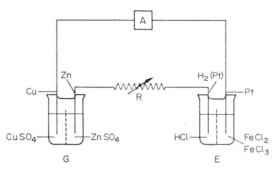

FIG. 1. The galvanic cell operates an electrolytic cell.

solution. In the compartment on the left the platinum electrode is surrounded with a tube through which hydrogen gas is bubbled, and is dipped into an acid solution. This works as a hydrogen electrode, the platinum again acting as an electron membrane. The electrode reaction on this electrode is

$$H_2 \text{ (gas on Pt)} = 2H^+ + 2e \qquad\qquad IV$$

The potential difference between the electrodes of this cell, when the concentrations of the solutions are one molar, is approximately 0·7 volts. Clearly, the first cell, having the larger voltage, will operate the second cell. (The terms voltage and potential are often used interchangeably.) Note the ammeter A for measuring the current I of this circuit and the variable resistor R. The latter can be used to vary the potential across the cell E: by increasing the resistance this potential will drop by IR and by decreasing the resistance the potential will rise until, when $R = 0$, the potential driving the reaction in E will be the potential across $G(V_G)$ minus that produced by $E(V_E)$.

$$V_G - V_E = IR \qquad\qquad 2.1$$

From the equation it is apparent that when $V_G < V_E$, I will reverse its sign, i.e. cell E will operate as a galvanic and cell G as an electrolytic cell. The dependence of cell voltage on concentration, i.e. the conditions under which $V_G < V_E$, are given by the Nernst equation, which will be dealt with later in this chapter.

Since the anode is the oxidizing electrode, the direction of the current through it (which by definition is opposite to the direction of flow of electrons) is from the electrode to the solution. This is the positive current. A cathodic or reducing current, which is negative, flows from the solution to the electrode. Thus, the identity of the electrodes in a cell is determined solely by the direction of the current; for a cell at equilibrium (i.e. zero current) the cathode and anode cannot be defined.

Any cell in operation must have an anode and a cathode. When we are interested in the reaction on one of these, we call the interesting electrode the "working" or "indicating" electrode. The other electrode is called the "counter" or "auxiliary" electrode. The name "reference electrode" is reserved for those electrodes which keep a constant potential against the bulk solution and are used in cells either as the counter electrode or as a third, measuring, electrode.

B. The structure of the interphase

An electrode reaction takes place at the surface of the electrode which is invariably in contact with the surface of the solution. This region, where the electrode and the solution meet and where their properties differ from those in the bulk is called the interphase and is the site of electrode reactions.

Why is the structure of the interphase different from the structure in the bulk? This can be answered by referring to Fig. 2, which presents the coulombic forces inside an ionic crystal, placed in vacuum. The forces on an ion in the bulk and on an ion on the surface are indicated by arrows: attractive forces are shown by arrows pointing from the central ion and repulsive ones are shown by arrows pointing toward the central ion. One sees that the forces acting on an ion in the bulk are symmetrical. They push or pull in all directions. When the ion is on the surface these forces work in half the directions. This means that the surface layer of a solid *must* be structurally different from the "inside" layers in order to cope with this change in forces. The same holds for species on the surface of a liquid. The forces, which in the bulk work in all directions, here work in half the directions only, changing the structure of the outside layer. That the structure of the surface of a liquid is different from that of the bulk is demonstrated by the existence of surface tension: the formation of drops and the behaviour of liquids in capillaries.

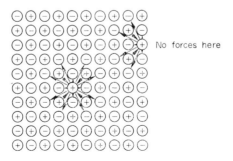

No forces here

FIG. 2. The coulombic forces on ions in the bulk and on the surface of an ionic crystal.

When a piece of metal is put into a polar liquid, two processes take place: (*a*) the dipoles of the liquid orient themselves toward the metal in a way which is a compromise between the force of the free electrons in the metal attracting the positive end of the dipoles and the force of the metal cations attracting the negative side of the dipoles; (*b*) some metal ions will dissolve, leaving a negative charge on the metal. This charge will polarize the surface layer of the solvent and the cations in solution will polarize the surface layer of the metal. The actual results differ from one system to another, but in principle what happens is that the surface layer of each phase acquires a charge which is equal in magnitude and opposite in sign for the two phases. Thus when one thinks of the interphase region, one must conclude that this region as a whole is neutral. Moreover, since there is charge separation between the electrode and the solution and the distance between the two charges is very small, one could think of the interphase as of a parallel-plate condenser. The concepts of capacity and potential difference across this condenser follow. The region of the interphase is very commonly called "the double layer" because of this analogy. Thus the charge on the electrode q is related to the capacity of the double layer C and to the potential difference across it ϕ by the well known formula

$$\phi = q/C \qquad\qquad 2.2$$

When solute molecules and ions are introduced into the polar liquid, a more complicated situation is encountered. The solute molecules will be either attracted or repulsed when reaching the interphase. When the solute consists of large molecules or ions, they will also be polarized in the electric field of the interphase. Thus the structure of the solution will be modified in the interphase region and some species may become adsorbed on the electrode as a result of this modification. Can this adsorption be treated in the same way as adsorption of gases on metals? There is one important difference between gas adsorption and solute adsorption on metals—in the gas-metal case the reaction is

$$\text{A (gas)} + \text{site} = \text{A (adsorbed)}, \qquad\qquad \text{V}$$

in the solute-solvent case the simplest reaction is

$$\text{A (solution)} + n\text{S (adsorbed)} = \text{A (adsorbed)} + n\text{S (solution)} \qquad \text{VI}$$

where S is a solvent molecule and n is the number of solvent molecules displaced by the solute. n may be smaller, equal or greater than one. Thus when solute molecules adsorb on the metal they must displace solvent molecules. The energy of adsorption will be different in these two cases, so will the adsorption isotherm.

A piece of ordinary solid metal is not a single crystal of metal. One always finds the grain boundaries where the forces, even inside the metal, are not spherically symmetrical. In addition, one always finds various kinds of imperfections inside a crystal where the forces are, again, assymetric. These faults are transferred from the bulk to the surface making the surface far from ideally smooth. Moreover, the surface of a solid can never be accurately represented by a continuous plane of atoms, the way it is often done. All these deviations from ideality give rise to various effects which do not represent the major phenomenon, but must be considered in the theoretical interpretation of experiments. These "side effects" are often so large as to mask the major phenomenon itself. A liquid metal is, of course, free from those side effects; it is smooth and not as structured as a solid one. Therefore it is more suitable for the study of the interphase itself without the side effects. Since kinetics of electrode reactions depend also on the state of the surface, liquid electrodes are also very suitable for kinetic studies. Thus one would use the liquid electrode (mercury) as often as possible in electrodics. Indeed, mercury occupies a unique position in electrodics because of the above reasons, as well as others which will become apparent later.

C. Mechanism of current flow

We have seen that the electrode-solution interphase acts like a parallel-plate condenser and that one could assign to it quantities that are commonly associated with a condenser, such as capacity, quantity of charge and potential difference. Taking the time derivatives of equation 2.2 yields

$$dq_c/dt = C d\phi/dt + \phi dC/dt \qquad\qquad 2.3$$

The subscript c was added to distinguish the charge of the capacitor from that of the electrode reaction.

$$dq_c/dt = I_c \qquad\qquad 2.4$$

which means that the current flowing through the circuit and charging the double layer is given by the change of potential with time plus the change of capacity with time. If the area of the electrode is constant, the time derivative of the capacity is zero, giving

$$I_c = C d\phi/dt \qquad\qquad 2.5$$

i.e. the current charging the double layer (which is also called capacity current) is proportional to the rate of change of potential with time. When direct current flows through the cell and there is no change of potential with time there is no capacity current. This means that direct current must be passed by another mechanism. Here we reach the basic principle of electrodics:

*The continuous flow of direct current through a cell is always associated
with two electrode reactions.*

One electrode reaction takes place at the cathode surface, the other at the
anode surface: once the current has crossed the electrode-solution interphase,
it is carried by ions through the solution until it reaches the other electrode-
solution interphase where it is transferred to the metallic conductor again.
This type of current is called "faradaic current".

For the reaction

$$Red = Ox + ne \qquad\qquad VII$$

one writes Faraday's law as

$$q_f = nFM \qquad\qquad 2.6$$

where q_f is the charge passed between the electrode and the solution and M is
the number of moles reacted with the charge q_f. Taking time derivatives

$$dq_f/dt = I_f = nF\,dM/dt \qquad\qquad 2.7$$

we find that the current is proportional to the rate of reaction of either
R or Ox.

The total current flowing through the cell is the capacity current *and* the
faradaic current. It is given by adding equations 2.5 and 2.7

$$I = I_c + I_f = nF\,dM/dt + C\,d\phi/dt \qquad\qquad 2.8$$

Thus, the current flowing through a cell is equal to the sum of two terms,
the first of which is proportional to the rate of the reaction and the second,
to the rate of change of potential.

At this point it is appropriate to say a few words on the conduction of
current through solutions. As is well known, the current is carried through
the solution by ions which differ from each other in their mobility. The
conductance of a solution (which is the reciprocal resistance) is proportional
to the number of ions in the solution and to some function of their mobility.
To find this relationship between conductance and mobility let us start with
Ohm's law

$$I = KV \qquad V = RI = \frac{1}{K}I \qquad 2.9$$

where K is the conductivity of the solution and V is the voltage across the
cell. When measuring the conductance, one likes to refer to a standard cell
of unit electrode area and unit length as the distance between the electrodes.
The "specific conductance" is

$$K = kA/l \qquad\qquad 2.10$$

K is the conductance measured in a rectangular cell where the area of the
electrodes is A and the distance between them is l; k is the specific conductance.
Consider a solution of concentration c. Imagine now that we measure the

conductance of this solution in a cell where $l = 1$ and the solution, whatever its volume, contains exactly one mole of the solute. Since the volume of solution of concentration c, which contains one mole, is $1/c$, and since in our special measuring cell $l = 1$, it follows that $A = 1/c$. The conductance in such a special cell is called "molar conductance" Λ

$$\Lambda = K/c \qquad 2.11$$

(c is given in moles m^{-3}), and

$$K = \Lambda c \qquad 2.12$$

The molar conductance is related to the number of ions in solution and to their mobility in the following manner

$$\Lambda = \sum_i n_i u_i z_i \varepsilon \qquad 2.13$$

where n is the number of ions in the cell, u is their mobility†, z is the charge of one ion in units of electronic charge and ε is the charge of an electron given in coulombs. Since the number of ions in the cell where Λ is defined is constant and equals Avogadro's number times the degree of dissociation (α) times the number of ions in each molecule (v), one writes

$$\Lambda = \alpha \varepsilon N \sum_i v_i u_i z_i = \alpha F \sum_i v_i u_i z_i \qquad 2.14$$

When there are several salts in solution, the specific conductance of the solution is the sum of the specific conductance of all salts. Thus

$$I = KV = V \sum_j \Lambda_j c_j = VF \sum_{ij} \alpha_j v_i u_i z_i c_i/v_i = VF \sum_{ij} \alpha_j u_i z_i c_i \qquad 2.15$$

where c denotes the concentration of the *ion* i in moles m^{-3}.‡ When current flows through the solution, different proportions of it are carried by different ions. These proportions are called the "transport numbers" (t); they are also sometimes called "transference numbers"

$$t_i = \frac{\alpha_j u_i z_i c_i}{\sum\limits_{ij} \alpha_j u_{.} z_i c_i}$$

and

$$\sum_i t_i = 1.$$

† The mobility of an ion is the velocity of its migration in a field strength of one volt.

‡ This is the simplest treatment of ionic conductance in solution. It is only strictly correct for $v_i = 1$ and where α is constant, i.e. the degree of dissociation of one salt is independent of the concentration of the other salts.

Summing up we conclude that there are two ways by which current can flow through the electrode solution interphase: by charging the double layer and by reacting with components in the solution. Current flows through the solution only by ionic migration, each ion sharing according to its mobility, charge and concentration.

D. Cell and electrode potential

The potential at a point x is defined as the amount of work needed to bring a unit charge from infinity to that point. This definition of electric potential is analogous to that of potential in magnetic or gravity fields. The above definition implies that we have arbitrarily defined the potential at infinity as zero and shows that the definition has meaning only when we speak of a *difference*, i.e. the amount of work needed to transfer a unit charge from one point (infinity) to another (x).

When dealing with potentials of electrodes and cells we can define a cell potential as the amount of work needed to transfer a unit charge from one electrode to the other. Let us now deal with the ways by which these cell potentials can be measured. Consider first the measurement of cell potential using Fig. 3. Here there are two electrodes M_1 and M_2 dipping into the solutions s_1 and s_2, respectively. The electrodes are connected through copper wires Cu to a resistance R. By a suitable calorimetric device the amount of heat generated at R in a unit time can be measured.

$$\text{Heat} = I^2 R \qquad\qquad 2.17$$

The current I could be measured by a silver coulometer, and the potential of the cell calculated from

$$V = IR \qquad\qquad 2.18$$

Thus the amount of work that was performed by transporting electrons from

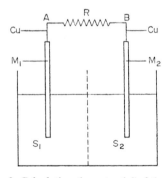

FIG. 3. Calculating the potential of the cell.

one electrode to the other as electric current, was determined as heat and was translated to potential. The relationship between cell potential and energy (heat or work) is thus established. It is also established that cell potential is a measurable quantity. This is not the most practical way, however, of measuring cell potential; other methods which use voltmeters or potentiometers drain very little current from the cell.

Let us now investigate the meaning of the cell potentials. Denote as $^2\phi^1$ the potential difference between phase 2 and phase 1 and consider Fig. 3; the sought potential, the work needed to transport a unit charge from A to B, is the sum of the work in transferring the charge through all phases and phase boundaries. Neglecting the resistance of the metals and assuming that the cell is at a constant temperature, we obtain

$$^A\phi^B = {}^{Cu}\phi^{M_1} + {}^{M_1}\phi^{s_1} + IR_{s_1} + {}^{s_1}\phi^{s_2} + IR_{s_2} + {}^{s_2}\phi^{M_2} + {}^{M_2}\phi^{Cu} \qquad 2.19$$

The potential difference between two metals (such as M_1 and Cu) is equal to the work needed to take a unit charge from one metal to infinity and bring it into the other metal. This amount of work is the difference between the work functions (W) of the electrons in the metals

$$^{Cu}\phi^{M_1} + {}^{M_2}\phi^{Cu} = W_{Cu} - W_{M_1} + W_{M_2} - W_{Cu} = \left(W_{M_1} - W_{M_2}\right) = {}^{M_2}\phi^{M_1} \qquad 2.20$$

Equation 2.20 shows that the material from which the measuring instrument is made does not influence the measured potential as long as the two leads are of the same material. Rewriting of equation 2.19, remembering that actual measurements are made at virtually zero current, gives

$$^A\phi^B = {}^{M_2}\phi^{M_1} + {}^{s_1}\phi^{s_2} + {}^{M_1}\phi^{s_1} + {}^{s_2}\phi^{M_2} \qquad 2.21$$

This means that the potential between two electrodes is equal to the sum of the potentials at the phase boundaries excluding the measuring instrument.

We next consider the potential of a single electrode. What is meant by "the potential of a single electrode"? We have so far considered potentials at points referred to infinity, but those who study electrode processes are interested in the energy required to transfer charge from the electrode not to infinity but to the solution. Let us assume that the potential in the bulk of the solution, ϕ_b, is constant, and call "electrode potential" the work required to transfer a unit positive charge from the solution to the electrode. Let us now go through the same argument for a single electrode; in Fig. 3 remove M_2 and s_2 and insert one of the leads of the measuring instrument into s_1, while the other remains attached to M_1. It is left to the reader to evaluate this new situation but the result is that the lead inserted into s_1 creates another metal-solution boundary with a potential difference $^{s_1}\phi^{Cu}$. The argument for equation 2.20 when applied to the present case is not useful since the properties of the measuring instrument do not cancel from the equation. If the properties of the measuring instrument enter a measured quantity, it means that with different instruments we shall obtain different results; in other words, the measurement has no meaning. *The potential difference between an electrode and its solution cannot be measured.*

Another way of looking at the same problem is via thermodynamics. The potential is related to the free energy of the corresponding reaction by equation 1.1. Therefore, in order to find the potential of a single electrode, we must find the free energy change for the reaction

$$A + e = A^-. \qquad \qquad \text{VIII}$$

ΔG for this reaction is

$$\Delta G = \int dG = \sum_{i,\,j} \left[\int B_i \, dD_i + \int \mu_j \, dn_j \right] \qquad 2.23$$

where B_i represents any extensive property such as volume, entropy, electrical charge etc., and D_i represents the corresponding intensive property such as pressure, temperature and electric field, respectively. The sum over i is over the properties. μ_j is the chemical potential of the species in the system, n_j is the number of moles of these species. The sum over j is over the various species in the system. When calculating the free energy of this reaction, we must be able to hold all variables, except one, constant and integrate the resulting equation

$$\Delta G = \int \mu_{A^-} dn_{A^-}, \quad dD_i = dn_{j \neq A^-} = 0$$

This is impossible to do with any charged species, since if charged particles are moved in or out of the system, the electric field in the system changes and the condition $dD_i = 0$ no longer holds. We cannot, therefore, calculate the free energy change of a reaction where charged particles (of only one sign) are involved, nor can we determine the potential of such an electrode.

If, however, we try to look for a scale of *relative* instead of *absolute* electrode potentials, we might be in a better position.

We shall leave the problem of electrode potential at this point in order to clarify certain terms. We defined the term "electrode potential" as the difference between the potential inside the metal electrode and the potential at the bulk of the solution. This means that the word "electrode" is no longer used to mean a piece of metal only but now includes the solution as well. Of course, an electrode potential has no meaning unless we consider the solution. This ambiguity may be removed by calling the combination metal-solution a "half-cell" and referring to "half-cell potentials", reserving the word "electrode" for the metal itself. However, this is not a universally accepted terminology and the reader must judge from the context the exact meaning of the word "electrode". Another, quite different definition for "electrode potential" will follow when we resume our discussion of relative potentials.

It was shown that cell potentials are measurable quantities. Consider the cell where the following reaction takes place

$$A + C = B + D \qquad \text{IX}$$

The potential across the terminals of this cell V is given by the Nernst equation

$$V = V^0 + \frac{RT}{nF} \ln \frac{a_B a_D}{a_A a_C} \qquad 2.22$$

where V^0 is the standard cell potential ($V^0 = -(RT/nF)\ln K$, K being the equilibrium constant for the reaction) and a denotes the activity of the respective species in solution. Equation 2.22 can be written in a different way

$$V = E^0{}_1 - E^0{}_2 + \frac{RT}{nF} \ln \frac{a_B}{a_A} - \frac{RT}{nF} \ln \frac{a_D}{a_C}$$

$$= \left(E^0{}_1 + \frac{RT}{nF} \ln \frac{a_B}{a_A} \right) - \left(E^0{}_2 + \frac{RT}{nF} \ln \frac{a_D}{a_C} \right) \qquad 2.23$$

which divides the cell potential into two "electrode potentials" whose respective reactions are

$$A = B + ne \qquad \text{IXa}$$

$$C + ne = D \qquad \text{IXb}$$

If we could find one electrode whose E^0 is very reproducible and where the activities could be made equal to one, we could use this electrode as a reference, measure all other electrodes against it and use this relative electrode potential scale. It was found that the hydrogen electrode is very suitable for this purpose. Thus the standard hydrogen electrode is defined as the electrode where the activity of the acid solution used is 1 (at the used concentration scale, here molar) and the pressure of the hydrogen gas (strictly speaking, the fugacity) is one atmosphere. The potential of this electrode is defined as zero at all temperatures.

$$E^0{}_{H_2/H^+} = 0$$

Therefore, when tables of electrode potentials are used, one should always remember that the values given are not $^M\phi^s$, but the potential of the cell:

$$\text{Pt, } H_2 \text{ (pressure 1 atm.)} \quad H^+(a = 1) \quad s \quad M \qquad \qquad X$$

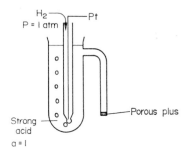

FIG. 4. A standard hydrogen electrode.

The construction of the standard hydrogen electrode is given in Fig. 4.

Let us now discuss the subject of sign of standard electrode potentials. The modern convention is that when saying "standard electrode potential of M^+/M" we mean the standard potential of the cell where the reaction

$$\tfrac{1}{2}H_2 + M^+ = H^+ + M \qquad\qquad XI$$

takes place. From equation 1.1 follows that if the reaction proceeds spontaneously, under standard conditions, from left to right the sign of E^0 is positive. Since the metal in reaction XI is being reduced, one will often find that these potentials are called "reduction" potentials and are listed in Tables with the half-reaction $M^+ + e = M$. Table I lists several standard electrode potentials with their two alternative presentations.

TABLE I. *Standard electrode potentials*

Electrode	Electrode reaction	E^0 (volts)
Li^+/Li	$Li^+ + e = Li$	$-3{\cdot}025$
K^+/K	$K^+ + e = K$	$-2{\cdot}925$
Na^+/Na	$Na^+ + e = Na$	$-2{\cdot}714$
Mg^{2+}/Mg	$Mg^{2+} + 2e = Mg$	$-2{\cdot}37$
Be^{2+}/Be	$Be^{2+} + 2e = Be$	$-1{\cdot}69$
Zn^{2+}/Zn	$Zn^{2+} + 2e = Zn$	$-1{\cdot}763$
Cd^{2+}/Cd	$Cd^{2+} + 2e = Cd$	$-0{\cdot}403$
$H^+/H_2, Pt$	$H^+ + e = \tfrac{1}{2}H_2$	$0{\cdot}000$
$Cl^-/AgCl, Ag$	$AgCl + e = Ag + Cl^-$	$+0{\cdot}2225$
$Cl^-/Hg_2Cl_2, Hg$	$\tfrac{1}{2}Hg_2Cl_2 + e = Hg + Cl^-$	$+0{\cdot}2675$
$Fe^{2+}, Fe^{3+}/Pt$	$Fe^{3+} + e = Fe^{2+}$	$+0{\cdot}771$
Ag^+/Ag	$Ag^+ + e = Ag$	$+0{\cdot}799$
Cl^-/Cl_2	$\tfrac{1}{2}Cl_2 + e = Cl^-$	$+1{\cdot}359$

One may find other conventions, such as "European" or "American" used in older publications. Therefore, caution must be exercised when reading such Tables in order to make sure that the signs are correctly used.

Since the use of the standard hydrogen electrode may be dangerous and is not convenient, other reference electrodes are usually used. The most common are the saturated calomel electrode, Cl^-/Hg_2Cl_2, Hg (in saturated KCl), the silver chloride electrode, $Cl^-/AgCl$, Ag and other silver halide electrodes. In order to be practical, a reference electrode should acquire its equilibrium potential rapidly and maintain it over long periods of time even when small currents are passing through it. This implies that the electrode reaction should be extremely rapid and reversible. Another term often used for an ideal reference electrode is "ideally non-polarizable electrode". An electrode which fulfils the requirements for a good reference electrode can have only one potential at zero current, regardless of its size or details of construction. It becomes an ideal non-polarizable electrode if it does not change its potential even when current flows through it. Thus, if we plot the current flowing through this ideal electrode as a function of its potential we get a vertical line (Fig. 5). This electrode has zero resistance.

At the other extreme one finds the "ideally polarizable electrode". This electrode does not transfer current, no matter what potential is applied to it. Its current-potential curve is a horizontal line and its resistance is infinity.

In practice there are no ideal electrodes, but one can construct one so as to approach ideality over certain regions of potential and current. The actual construction and properties of various electrodes will be discussed in Chapter 5.

E. Overpotential

The Nernst equation gives values for electrode potentials at equilibrium, i.e. those measured when the electrochemical potential of the reactants (including the electrons) is equal to that of the products. There is no net

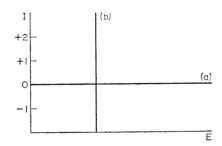

FIG. 5. Current potential curves for ideal (a) polarizable, (b) non-polarizable electrodes.

reaction at equilibrium. What happens when an electrode reaction takes place? The electrochemical potentials of the reactants and products are no longer equal. This means that the potentials of the electrodes considered will no longer be that of equilibrium but some other ones.

We define the overpotential η as the difference between the actual potential of the working electrode E and the equilibrium potential of that electrode E^e.

$$\eta = E - E^e \qquad\qquad 2.24$$

It is convenient to assign names to measured overpotentials according to the mechanism by which the reaction proceeds. Generally an electrode reaction may proceed through some of the following steps:

(1) the electroactive particle is transported to the electrode surface, *or*

(2) the electroactive particle is created by a chemical reaction from non-electroactive particles;

(3) the electroactive particle is adsorbed on the electrode surface;

(4) electrons are transferred between the electrode and the reactant;

(5) the reacted particle is transported to the bulk of the solution, *or*

(6) the reacted particle is first desorbed and then undergoes step 5, *or*

(7) the reacted particle reacts chemically, producing either electroactive or non-electroactive products, *or*

(8) the reacted species is incorporated onto the electrode surface as in crystallization of metals, oxides etc.

The term "electroactive particle" was used here to indicated the particle which reacts at the electrode at the given potential. It must not be identical with the reactant introduced into the solution if step (2) takes place.

We see that, in spite of the great complexity of electrode reactions, we can divide the steps into three general categories: (*a*) concentration, (*b*) surface and (*c*) electron transfer. The first category includes all mechanisms in which the supply of the reactant is the rate limiting step. The overpotential associated with this kind of mechanism is called "concentration overpotential" (or, sometimes, "concentration polarization"). The second category includes all reactions for which either adsorption, desorption or crystallization influence the rate determining step. The overpotential associated with these processes is called "adsorption" overpotential or "crystallization" overpotential. The third category includes those processes for which the electron transfer is the rate determining step and the corresponding overpotential is named accordingly "electron transfer overpotential".

3. Electrode Kinetics

As the name of this chapter implies, we shall deal here with the various rate expressions associated with different types of electrode reaction mechanisms. But first let us compare electrode kinetics with "regular" solution kinetics and evaluate the similarities and the differences between the two.

If we think of the general reaction

$$aA + bB = cC \qquad\qquad I$$

we can write the rate expression as

$$v = \vec{k}[A]^{\vec{\alpha}}\,[B]^{\vec{\beta}}\,[C]^{\vec{\gamma}} - \overleftarrow{k}[A]^{\overleftarrow{\alpha}}\,[B]^{\overleftarrow{\beta}}\,[C]^{\overleftarrow{\gamma}} \qquad\qquad 3.1$$

where $\vec{k}$ and $\overleftarrow{k}$ are the forward and reverse rate constants, $\vec{\alpha}$, $\vec{\beta}$, $\vec{\gamma}$ and $\overleftarrow{\alpha}$, $\overleftarrow{\beta}$, $\overleftarrow{\gamma}$ are the forward and reverse orders of the reaction and the square brackets represent concentrations.†

It must be remembered that a general reaction cannot exclude the possibility of autocatalysis or inhibition by products, in which case $\vec{\gamma}$ or $\overleftarrow{\alpha}$ or $\overleftarrow{\beta} \neq 0$. If, for example B is replaced with an electron, we are immediately in the realm of electrode processes, because only in this area can we speak meaningfully on reactions with electrons. The corresponding rate equation for the reaction

$$aA + ne = cC \qquad\qquad II$$

is

$$v = \vec{k}[A]^{\vec{\alpha}}\,[e]^{\vec{\beta}}\,[C]^{\vec{\gamma}} - \overleftarrow{k}[A]^{\overleftarrow{\alpha}}\,[e]^{\overleftarrow{\beta}}\,[C]^{\overleftarrow{\gamma}} \qquad\qquad 3.2$$

In equation 3.1 the meaning of all terms was quite clear; in equation 3.2, the "concentration" of electrons needs explanation. In order to do so, we shall discuss energy and concentration.

Both the chemical potential of a substance i, μ_i, and the potential at a point are quantities of energy. μ_i is the change in Gibbs free energy of the system when an infinitesimal amount of material i is transferred into it from the standard state of i (keeping the temperature, pressure and concentrations of the other components constant); the potential, on the other hand, is the

---

† Opinions about the use of concentrations or activities in rate expressions vary. In this monograph concentrations will be used throughout because it is the quantity which is available experimentally.

amount of energy needed to bring a unit charge from infinity to the point in question. (If that point is inside a material phase than the composition of the material, as well as the temperature and pressure, must remain constant.) In order to be able to compare chemical and electrical potentials, the amounts of material and charge involved must be equalized. The amount of work needed to bring a charge ε through potential difference ϕ, is $\varepsilon\phi$; ε may, of course be infinitesimal. Thus we may compare μ_i and $\varepsilon\phi$. If we consider, as we do, systems of solutions in contact with electrodes, it is quite clear that the energy of such systems can be raised either by adding material i or by bringing electrical charge to them. Thus the chemical potential and the electrical energy ($\varepsilon\phi$) can be treated as analogous quantities.

As a first approximation, the relation between chemical potential and concentration of species A is

$$\mu_A = \mu_A{}^0 + RT \ln \frac{[A]}{[A]_{ss}} \qquad 3.3$$

where $[A]_{ss}$ is the concentration of A at the standard state, taken as 1 molal. Likewise one could write an analogous equation for the relation between the electrical energy and the "concentration" of the electrons, remembering that one mole of electrons is F (F is the Faraday) coulombs.

$$F\mathscr{E} = F\mathscr{E}^0 + RT \ln \frac{[e]}{[e]_{ss}} \qquad 3.4$$

I used $\mathscr{E}$ here instead of ϕ because this equation was not derived rigorously: the standard state of electrons cannot be defined in terms of concentration and therefore $\mathscr{E}^0$ cannot be defined in these terms either. But we may write $\mathscr{E} - \mathscr{E}^0 = \phi$, i.e. the potential of the electrode $\mathscr{E}$ minus the potential at the standard state is the electrode potential ϕ. Thus

$$n F (\mathscr{E} - \mathscr{E}^0) = n F \phi = RT \ln \frac{[e]}{[e]_{ss}} \qquad 3.5$$

and therefore

$$[e] = [e]_{ss} \exp \frac{n F}{RT} \phi \qquad 3.6$$

The rate expression now becomes

$$v = \vec{k}[A]^{\bar{x}} [C]^{\bar{y}} [e]_{ss}^{\bar{\beta}} \exp \frac{\vec{\beta} n F}{RT} \phi - \overleftarrow{k}[A]^{\bar{x}} [C]^{\bar{y}} [e]_{ss}^{\beta} \exp \frac{\overleftarrow{\beta} n F}{RT} \phi \qquad 3.7$$

Since $[e]_{ss}$ is constant for the reaction, it may be included in k

$$\left.\begin{aligned}
v &= \vec{k}[A]^{\bar{z}} [C]^{\bar{y}} \exp \frac{\vec{\beta}nF}{RT} \phi - \overleftarrow{k}[A]^{\bar{z}} [C]^{\bar{y}} \exp \frac{\overleftarrow{\beta}nF}{RT} \phi \\
\vec{k} &= \vec{k}[e]_{ss} \\
\overleftarrow{k} &= \overleftarrow{k}[e]_{ss}
\end{aligned}\right\} \quad 3.8$$

The rate of reaction I is defined as the rate of change of amount of material with time divided by the number of moles of the substance for the formula as written

$$v = \frac{1}{a}\frac{d(A)}{dt} = \frac{1}{b}\frac{d(B)}{dt} = \frac{1}{c}\frac{d(C)}{dt} \qquad 3.9$$

the parentheses represent amount. Likewise, the rate of electrode process is

$$v = \frac{1}{a}\frac{d(A)}{dt} = \frac{1}{n}\frac{d(e)}{dt} = \frac{1}{c}\frac{d(C)}{dt} \qquad 3.10$$

The "amount" of electrons must, on the one hand be expressed in electrical charge, i.e. coulombs; on the other hand it must be, like all other terms in equation 3.9, in units of moles sec^{-1}. Obviously we need to introduce a factor to convert coulombs to moles. Such a factor is the Faraday, whose units are coulombs $mole^{-1}$. Therefore

$$\frac{1}{n}\frac{d(e)}{dt} = \frac{1}{nF}\frac{dq}{dt} = \frac{1}{nF}I \qquad 3.11$$

q is the charge passed through the electrode; dq/dt is the current, I, passed in the reaction. Thus the rate of electrode reaction is directly proportional to the current and depends on the exponent of the potential of the electrode.

Before we start the detailed discussion of electrode kinetics let us look at the dimensions of the reaction rate constants for solution reactions as compared to that of electrode reactions. In solution kinetics the important factors to be considered are the concentrations of reactants and the volume of the reaction mixture. If the volume, V, is constant, we can obtain a "reaction rate density" by dividing v by V

$$\frac{v}{V} = \frac{1}{aV}\frac{d(A)}{dt} = \frac{1}{a}\frac{d(A)/V}{dt} = \frac{1}{a}\frac{d[A]}{dt} \qquad 3.12$$

This reaction rate density is the rate of change of *concentration* (as opposed to amount) with time. However, the term is not generally used and "reaction rate" is used for both amount rates and concentration rates. In electrode

kinetics the important factors are concentrations of reactants and area of
the electrode; therefore the meaningful rate of reaction is given by the current
density, i. Dimensional analysis of equation 3.1 (taking v as the reaction rate
density) shows that the dimensions of k are $[\sec^{-1} \text{moles}^{-(\alpha+\beta+\gamma-1)}$
$\text{cm}^{3(\alpha+\beta+\gamma-1)}]$. Dimensional analysis of

$$\frac{i}{nF} = \vec{k}[A]^{\bar{\alpha}} [C]^{\bar{\gamma}} \exp \frac{\bar{\beta}nF}{RT} \phi - \overleftarrow{k}[A]^{\bar{\alpha}} [C]^{\bar{\gamma}} \exp \frac{\bar{\beta}nF}{RT} \phi \qquad 3.13$$

which is a combination of equations 3.8 and 3.11 (introducing $i = I/A$),
shows that the dimensions of k are $[\sec^{-1} \text{moles}^{-(\alpha+\gamma-1)} \text{cm}^{3(\alpha+\gamma-1)} \text{cm}]$.
Thus the reaction rate constants of solution and electrode reactions differ in
their dimensions. This implies that there is a basic difference between them,
the difference between volume and surface reactions.

A. The passing of a Faradaic current—a surface process

It has already been established that in order to enable electron transfer to
take place, the electro-active species must first reach the electrode surface
and, often, be adsorbed on to it. In this section, we deal first with the laws
governing mass transport and then with the basic concepts of adsorption.

1. The laws of mass transport to the electrode surface.

Let us consider reaction III, proceeding from left to right

$$\text{Red} = \text{Ox} + n\text{e} \qquad \qquad \text{III}$$

There are three ways, or modes, by which Red can reach the electrode
surface: (a) by migration in the electrode's electric field, if it carries a charge;
(b) by the convective movement of the solution and (c) by diffusion in a
concentration gradient, the latter being produced by the depletion of
Red molecules at the surface due to the electrode reaction. Sometimes the
reactant is present in the electrode itself, such as when Red is the metal and
Ox is the metal ion.

The question to be dealt with now is how to measure mass transport. This
is done by measuring the amount of material passed through some place
during a certain length of time. In Fig. 6 the electrode is shown to lie in
the $y-z$ plane and the line of mass transport is along the x axis. The flux
(which is the quantity to be measured) is that number of moles of material
which pass during unit time (second) through a unit area (cm^2) parallel to
the $y-z$ plane, in the x direction. The flux may depend on all coordinates,
x, y and z, but in practice it is convenient to assume that all mass transport

is in the $-x$ direction only and is independent of y and z. This uni-dimensional mass transport is called "infinite" or "semi-infinite linear mass transport". The reason for this crude simplification is that the mathematics for infinite linear mass transport is fairly simple, while that for the real situation is very

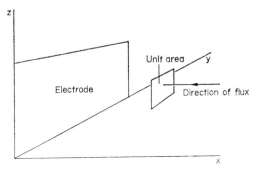

FIG. 6. The flux of material.

complicated. The justification for this simplification is that the laws obtained using it describe the experimental results fairly well.

Let us now discuss the relationship between the flux and the energies involved with the three modes of mass transport. Energy and work are equivalent terms—when the system loses energy it is doing work, when it gains energy, work is done on it. Moreover, energy and work are equal in value; this value of energy is given by the force doing the work times the distance over which the work is done

$$dW = Fdx \qquad 3.14$$

where F is the operating force and dW is the work (energy). At constant temperature and pressure, changes in energy of a system can come about because of changes in chemical composition and of electrical field. If we take one gram-mole of substance i from point 1 where the chemical potential is μ_1 and the electrical potential is ϕ_1, to a point 2 where these values are μ_2 and ϕ_2, the amount of work done on i is

$$W = \mu_2 - \mu_1 + zF\phi_2 - zF\phi_1 = \Delta\mu + zF\Delta\phi$$
$$= (\mu_2 + zF\phi_2) - (\mu_1 + zF\phi_1) = \bar{\mu}_2 - \bar{\mu}_1 \qquad 3.15$$

z is the ionic charge. Equation 3.15 defines a new quantity, the electro-chemical potential $\bar{\mu}$. By combining equations 3.14 and 3.15 we see that the force operating on a system subject to concentration gradient and electric field is

$$F = \frac{d\bar{\mu}}{dx} = \frac{d\mu}{dx} + zF\frac{d\phi}{dx} \qquad 3.16$$

The flux of material, say Red, must be proportional to the concentration of Red and to the operating force; the proportionality constant is D_R/RT, where D_R is the diffusion coefficient of Red. Sometimes, such as in flowing solutions, there is a flux without the existence of force. Then the flux is given by the product of the concentration and the velocity by which it flows, i.e. the velocity of flow of the solution. In equation form:

$$J = -c\left(\frac{D_R}{RT}\frac{d\mu}{dx} + \frac{D_R zF}{RT}\frac{d\phi}{dx} - v\right) \qquad 3.17$$

where J denotes the flux and v the velocity in the direction perpendicular to the electrode. The signs of the terms containing gradients of chemical and electrical potential are negative because Red is moving *toward* the electrode, i.e. in the direction from large x to small x, which is the negative direction. The sign of cv is positive because v is the component of velocity in the direction of flux. Taking the derivative of μ and opening brackets gives

$$J = -D_R\frac{dc}{dx} - \frac{D_R zFc}{RT}\frac{d\phi}{dx} + cv \qquad 3.18$$

After various manipulations one obtains two important results. The first gives the change of concentration with time

$$\frac{\partial c}{\partial t} = D_R\frac{\partial^2 c}{\partial x^2} - u\frac{z}{|z|}\frac{d\phi}{dx}\frac{\partial c}{\partial x} - v\frac{\partial c}{\partial x} \qquad 3.19$$

where u is, as before, the mobility of the ion considered, $z/|z|$ is the sign of the ionic charge. The second result is an expression for u

$$u = \frac{D|z|F}{RT} \qquad 3.20$$

which enables the calculation of D from conductance measurements.

As will be shown in Chapter 4, the extent of the double layer, i.e. that region where a field exists, is quite small in concentrated solutions, so that for the greatest part $d\phi/dx = 0$. In many experiments it is possible to maintain a quiescent solution, so that $v = 0$. It is not possible, however, to eliminate the diffusion term; when current passes through a cell there are always concentration changes at the surface of the electrode, always a concentration gradient and always diffusion.

The equation generated by substituting $v = 0$ and $d\phi/dx = 0$ in equation 3.18 is called Fick's first law. The equivalent equation corresponding to 3.19

is Fick's second law.

$$J = -D\frac{dc}{dx} \qquad\qquad 3.21$$

$$\frac{\partial c}{\partial t} = D\frac{\partial^2 c}{\partial x^2} \qquad\qquad 3.22$$

2. *General treatment of adsorption of electroactive species*

When adsorption of reactants or products influences the electrode reaction, one wants to know the extent of this influence using quantitative methods. In order to quantify adsorption, let us start with the list of things which we know and might be useful and a list of things which we wish to know about adsorption. Then we shall proceed trying to correlate the two lists, making those assumptions which seem necessary and justified about the system.

We often know the nature of the absorbing species (absorbate) and its concentration in solution. We also know the potential of the electrode. Sometimes we can measure the surface excess Γ (a definition of which is given in Chapter 4) which is related to the surface concentration L. Two items make the list of the desired quantities or functions: how does the nature and concentration of adsorbate and the potential of the electrode influence L and what is the value of L? The function which relates L to solution concentration c, to the nature of the species and to potential is called adsorption isotherm

$$f(L) = g(\Delta G^0_{ads}, c, E) \qquad\qquad 3.23$$

The nature of the adsorbate is expressed in ΔG^0_{ads}, the free energy of adsorption.

Unfortunately, we know very little about adsorption and cannot derive the functions f and g for the general cases. It is here that we must make assumptions and simplifications. First, we will not consider the electrode potential E at this stage, but will deal with it again later in this chapter. The second simplification is the way in which adsorption isotherms are derived: we start with the simplest possible model and derive an isotherm for it. Later, by adding terms perhaps, we make the simple equation apply to more complicated situations. The method which we would like to use is, of course, the exact opposite—derive an isotherm for the most complicated system first and then simplify it for simpler cases. With our present knowledge, however, this is not practical and so we shall derive the classical Langmuir isotherm for the case of adsorption from the gas phase and then examine the influence of solvent on this isotherm for the case of adsorption from solution. Later we shall consider two other isotherms which are extensions of Langmuir's and which may be used for better fit of experimental data.

In order to derive Langmuir's isotherm let us consider the following reaction

$$A(gas) + site = A(adsorbed) \qquad\qquad IV$$

If every site reacts with one molecule A independently of the other sites, we may write the following equilibrium expression

$$K_{ads} = \frac{L_A}{L_{site}[A]} \qquad\qquad 3.24$$

If the concentration of sites (per m²) on the bare electrode is L_{max} (L_{max} is also the maximum concentration of A(adsorbed)), then $L_{site} = L_{max} - L_A$ and equation 3.24 becomes

$$K_{ads} = \frac{L_A}{[A](L_{max} - L_A)} \qquad\qquad 3.25$$

If we substitute $\theta = L_A/L_{max}$, we get immediately

$$K_{ads}[A] = \frac{\theta}{1 - \theta} \qquad\qquad 3.26$$

which is the Langmuir isotherm. However, when interested in adsorption from solutions, one must consider not reaction IV but the exchange reaction

$$A(solution) + n \; solvent(ads) = A(ads) + n \; solvent \qquad\qquad V$$

Since we deal with solutions, we may write the equilibrium constant expression in terms of mole fractions

$$K_{ads} = \frac{X_{A(ads)} X_s^{\,n}}{X_{A(sol)} X_{s\,(ads)}^{\,n}} \qquad\qquad 3.27$$

where the subscript s stands for solvent and (sol) means "in solution". In fairly dilute solutions

$$X_s = 1$$

$$X_{A(sol)} = [A]/[s]$$

$$X_{A(ads)} = \frac{L_A}{L_A + L_s}$$

$$X_{s(ads)} = \frac{L_s}{L_A + L_s}$$

$$L_A = L_{max}\,\theta$$

$$L_s = nL_{max}(1 - \theta)$$

Substituting all the above relations in equation 3.27 yields

$$K_{ads} \frac{[A]}{[s]} = \frac{\theta}{(1-\theta)^n} \frac{[\theta+n(1-\theta)]^{n-1}}{n^n}$$ 3.28

If the second term on the right-hand side does not change much with θ, then equation 3.28 is just a modified form of the Langmuir isotherm when every adsorbed molecule occupies independent n sites and the value of the equilibrium constant is changed.

$$K'_{ads}[A] = K_{ads} \frac{[A]}{[s]} \frac{n^n}{[\theta+n(1-\theta)]^{n-1}} = \frac{\theta}{(1-\theta)^n}$$ 3.29

when θ is very small equation 3.29 becomes

$$K_{ads} \frac{[A]}{[s]} n = \frac{\theta}{(1-\theta)^n} \sim \theta$$ 3.29a

when θ is very large equation 3.29 reads

$$K_{ads} \frac{[A]}{[s]} n^n = \frac{\theta}{(1-\theta)^n} \sim \frac{1}{(1-\theta)^n}$$ 3.29b

when n is not very large (in the order of 2, 3 or 4) the ratio of 3.29a to 3.29b is not very small. This means that the right-hand side of equation 3.29 does not change very much with θ.

Fig. 7 shows that in fairly low coverages, equation 3.28 gives essentially the same curve as the modified form of the simple Langmuir isotherm, equation 3.29. The curves in Fig. 7 were calculated for $n = 2$. This discussion is to emphasize that, quite often, we need not consider the difference between adsorption from the gas phase and adsorption from solution; the only significant difference is the value of the equilibrium constant.

It is known from thermodynamics that

$$\Delta G^0_{ads} = -RT \ln K_{ads}$$ 3.30

The assumption of Langmuir's isotherm, that all sites are independent of each other, implies that ΔG^0_{ads} does not change with coverage. Let us now see what happens when we let ΔG^0_{ads}, as obtained from the Langmuir isotherm, vary linearly with coverage, since in real systems the sites are not independent of each other; there is interaction between adsorbed molecules on neighbouring sites.

$$\Delta G^0_{ads} = \Delta G^0_{ads} + b\theta$$ 3.31

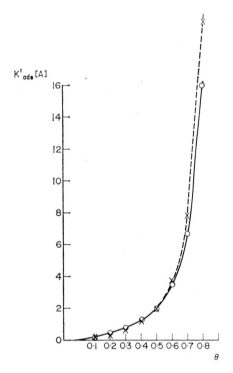

FIG. 7. The Langmuir adsorption isotherm for ^{o} adsorption from solution, calculated from equation 3.28, ^{x} adsorption from the gas phase, calculated from equation 3.29.

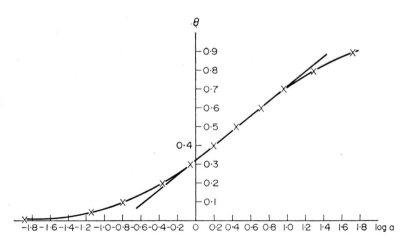

FIG. 8. The Frumkin adsorption isotherm $\Delta G^{o} = 566$ J mole^{-1}, $b = 4180$ J mole^{-1}.

Introducing equations 3.30 and 3.31 into 3.26 gives what is known as the Frumkin adsorption isotherm (Fig. 8)

$$\frac{\theta}{1-\theta} \exp \frac{b\theta}{RT} = K_{ads}[A] \qquad\qquad 3.32$$

$$\theta + \ln \frac{\theta}{1-\theta} = \frac{RT}{b} (\ln K_{ads} + \ln [A]) \qquad\qquad 3.33$$

where $K_{ads} = \exp - (\varDelta G^0_{ads}/RT)$. Since $\ln (\theta/1-\theta)$ is very small in the vicinity of 0·5 (such is the property of the ln function), a plot of θ as a function of log $[A]$ according to equation 3.32 shows that for coverages between 0·2 and 0·7 the curve is very nearly linear. This linear dependence of θ on log $[A]$ is known as the approximate form of Temkin's adsorption isotherm

$$\theta = \frac{2\cdot3RT}{b} \log K_{ads} + \frac{2\cdot3RT}{b} \log [A] = K + \frac{2\cdot3RT}{b} \log [A] \quad 3.34$$

The isotherm used in a particular case depends on the experimental data. It is chosen to give the best fit, with K and b as experimental parameters which can be adjusted.

Adsorption phenomena will be discussed again in Section F, when we will consider their direct influence on the kinetics of electrode reaction, and in Chapter 4, where the theory of the interphase will be developed.

B. The current–voltage curve

The commonest form of data on electrode processes is that of current-voltage curves. When there is an electrolytic cell made of one polarizable and one non-polarizable electrode and a potential difference is imposed between them, it can be safely assumed that the potential of the non-polarizable electrode remains constant, the imposed voltage changes the potential of the polarizable (working) electrode alone. Thus, when dealing with the reaction

$$\text{Red} = \text{Ox} + n\text{e} \qquad\qquad \text{VI}$$

we can start at a potential negative enough so that all electrons remain in the solution bound to Red, i.e. no reaction takes place. When the potential becomes more positive, a value will be reached where some electrons will leave Red and a small current will be observed. The current will increase as the potential becomes more positive until a point will be reached where Red is being oxidized so rapidly that the concentration of it near the surface is zero. The reaction rate is now equal to the rate of mass transport. When the only mode of mass transport is diffusion or controlled convection, the limiting current density is proportional to the bulk concentration of the electroactive species

$$i_{la} = \alpha_a [R]_0 \qquad\qquad 3.35$$

The current-voltage curve will have a shape resembling that of a titration curve (see Fig. 9a). When the wanted reaction is the reverse of VI, we can start at a very positive potential so that no reduction takes place and proceed to more negative potentials. Here, too, a cathodic limiting current will be observed at negative enough potentials

$$i_{lc} = \alpha_c [Ox]_0 \qquad\qquad 3.36$$

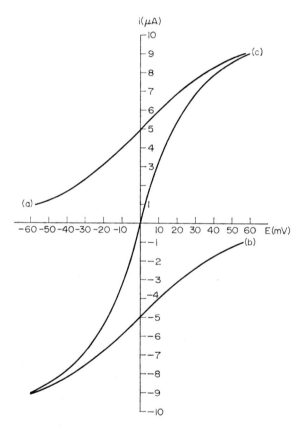

FIG. 9. Current-potential curve for reversible electrode processes (the rising portion of the wave).

(Fig. 9b). When reaction VI can take place in both directions then the current-voltage curve is as in Fig. 9c.

The term "reversible reaction" is often found in polarographic literature. This may have one of two meanings. It may mean, as in ordinary kinetics, a reaction which can be made to go in either direction, or it may mean that the reaction is both reversible (kinetically) and very, very rapid. The argument is as follows: when an electrode reaction is so rapid that the technique used for measurement is too slow to give kinetic information, the reaction is virtually at equilibrium and equilibrium methods, i.e. thermodynamics and the Nernst equation, can be used to describe it. Since the current at any potential is proportional to the difference in concentration between the bulk and the electrode surface, we can write

$$i_a = \alpha_a([R]_0 - [R]) = i_{la} - \alpha_a[R]$$ 3.37

$$i_c = \alpha_c([Ox]_0 - [Ox]) = i_{lc} - \alpha_c[Ox]$$ 3.38

where i_a and i_c stand for anodic (oxidation) and cathodic (reduction) current densities respectively. The subscript 0 denotes the respective concentration at the bulk but no subscript denotes the same near the electrode. From Nernst's equation

$$E = E^0 + \frac{RT}{nF} \ln \frac{[Ox]}{[R]}$$ 3.39

can be derived, using equations 3.37 and 3.38, the following expression for the current-voltage curve of a "reversible" reaction

$$E = E^0 + \frac{RT}{nF} \ln \frac{\alpha_a}{\alpha_c} + \frac{RT}{nF} \ln \frac{i_c - i_{lc}}{i_{la} - i_a}$$ 3.40

When diffusion only is the mode of mass transport, α_a and α_c can be evaluated by solving equation 3.22. It appears that α_a is proportional to $D_R^{\frac{1}{2}}$ and α_c to $D_{Ox}^{\frac{1}{2}}$ with the same proportionality constant. Therefore

$$E = E^0 + \frac{RT}{2nF} \ln \frac{D_R}{D_{Ox}} + \frac{RT}{nF} \ln \frac{i_c - i_{lc}}{i_{la} - i_a} = E_{\frac{1}{2}} + \frac{RT}{nF} \ln \frac{i_c - i_{lc}}{i_{la} - i_a}$$ 3.41

The curve in Fig. 9c was calculated using equation 3.41 assuming $E^0 = 0$, $D_R = D_{Ox}$, $n = 1$ and limiting currents of 10^{-5} A cm^{-2}. Compare this to the curve in Figs 13 and 14, which are for reactions which are kinetically reversible, but slow, so that Nernst's equation cannot be used for their description.

Let us now derive the equation for the curve in Fig. 9a. Start, as before, with the Nernst equation and remember that for every molecule Red oxidized, one Ox molecule is produced, so that

$$i_a = \alpha_a([R]_0 - [R]) = \alpha_c[Ox] \qquad 3.42$$

from which is obtained

$$E = E^0 + \frac{RT}{2nF} \ln \frac{D_R}{D_{Ox}} - \frac{RT}{nF} \ln \frac{i_{la} - i_a}{i_a} = E_{\frac{1}{2}} - \frac{RT}{nF} \ln \frac{i_{la} - i_a}{i_a} \qquad 3.43$$

A similar equation can be derived for a reduction process, such as in Fig. 9b.

Let us now turn to detailed discussion of the various types of overpotential. When doing so it is necessary to keep in mind the assumption that it is only the discussed step which limits the reaction rate, i.e. when discussing reaction overpotential we assume that all other steps, diffusion, electron-transfer etc., are virtually at equilibrium and derive the equations accordingly. In practice this assumption is not always correct. There may be two steps of the same rate, both of which limit the electrode reaction. For example the reduction of $Cd(CN)_4{}^{2-}$ at the dropping mercury electrode proceeds via the mechanism

$$Cd(CN)_4{}^{2-} \overset{fast}{\rightleftarrows} Cd(CN)_3{}^- \overset{slow}{\rightleftarrows} Cd(CN)_2 \overset{+2e}{\rightleftarrows} Cd(Hg) \qquad VII$$

When the overpotential is small, the rate determining step is the electron transfer, the kinetics of this step can be determined using appropriate techniques. When the overpotential increases, the rate limiting step becomes the formation of $Cd(CN)_2$. This rate, too, can be determined by experiment. In the middle potential range both rates are approximately equal and experimental data will be inconclusive. It often happens that by changing experimental conditions, such as concentration, potential, rate of change of potential or current, the rate determining step can be changed and thus both steps studied. Mathematical treatments of many complex electrode reactions are available in the literature. These complex situations will not be discussed here.

C. Mass transport overpotential

Let us assume that the electric field of the interphase is confined to a very small distance near the electrode and does not affect the mass transport process. Equation 3.19 then becomes

$$\frac{\partial c}{\partial t} = D \frac{\partial^2 c}{\partial x^2} - v \frac{\partial c}{\partial x} \qquad 3.44$$

In order to solve this equation, initial and boundary conditions must be specified. It is always true that the concentration of the electroactive species,

c, is uniform throughout the solution at the beginning of an experiment

$$\text{when } t = 0, c = c_0 \quad \text{for} \quad x > 0 \qquad 3.45$$

This condition does not hold at the electrode surface $x = 0$, since the ionic distribution in the double layer alters the concentrations there. When the amount of material reacted during one experiment is very small compared to the amount present in the cell, the concentration in the bulk of the solution does not change during the experiment

$$\text{for } t > 0, x \to \infty, c \to c_0 \qquad 3.46$$

When we are interested in the current as a function of time, another boundary condition is

$$i = nFD \left(\frac{\partial c}{\partial x} \right)_{x=0} \qquad 3.47$$

The above equation shows that the current is proportional to the material flux. This is quite evident from Faraday's law: the flux is the number of moles passing through a unit area of surface in a second. This number of moles is equal to $D(\partial c / \partial x)$ since at $x = 0$ there is no solution movement: $v = 0$. This quantity (the flux) reacts with nF coulombs per mole. The total charge density per second is, therefore, given by the above expression.

For quiescent solutions, $v = 0$, the above set of differential equation and boundary conditions can be solved analytically using the Laplace transform procedure. It is outside the scope of this book to go through the mathematics in detail and only the results will be given here. For reaction VI, the relation between the current density and time at a constant potential, i.e. where $[R](x = 0, t > 0) = $ constant, is

$$i = nF([R]_0 - [R])(D_R/\pi t)^{\frac{1}{2}} \qquad 3.48$$

or

$$i = nF([Ox] - [Ox]_0)(D_{Ox}/\pi t)^{\frac{1}{2}} \qquad 3.49$$

These equations show that current passed through an electrode whose potential is held constant and where diffusion is the only mode of mass transport, does not reach a steady value. Thus when an experiment is carried out under these conditions and a steady value of current is measured, it is a sure sign that convection is aiding diffusion. This convection, known as "natural convection" arises from the density gradient formed by large concentration gradients near the electrode.

The concept of diffusion layer is useful. If one assumes that there exists a layer near the electrode of thickness δ, so that

$$i = nF \, D_{Ox} \frac{[Ox] - [Ox]_0}{\delta_{Ox}} = nF \, D_R \frac{[R]_0 - [R]}{\delta_R} \qquad 3.50$$

it can be seen from equations 3.48 and 3.49 that δ does not reach a constant value in a perfectly quiescent solution. When the solution is stirred, δ does reach a constant value after a short time and a steady current is observed. Of particular interest among the cases of stirred solutions is that of the rotating disk electrode. The interest in this system is because the hydro-dynamic equations of motion can be fairly easily solved for this case on the assumption that the whole area of the electrode is uniformly accessible, i.e. the planes of constant concentration are parallel to the rotating disk electrode. Here δ is given by Levitch's equation

$$\delta = 1 \cdot 61 D^{1/3} \, v^{1/6} \, \omega^{-\frac{1}{2}} \qquad\qquad 3.51$$

where v is the kinematic viscosity (viscosity divided by density) and ω is the angular velocity of the disk in radians s^{-1}.

If the potential of the working electrode varies with time it can do so either independently or as the dependent variable. The potential can be varied independently by adjusting the setting of a potentiometer or by using an instrument which gives a known function of potential with time. The dependent quantity will often be the current passed. If, however, we control the current by a suitable device, keeping it constant or changing it as a known function of time, the potential then varies as a function of the current and is, therefore, the dependent variable.

If we want to solve the diffusion problem for electrodes where the potential varies with time, then we must add to equations 3.44–3.47 an expression which relates the electrode potential E to the concentration at the electrode surface. Since we are dealing in this section with mass-transport controlled processes, we shall use a modified form of the Nernst equation. At zero current the potential is given by

$$E^e = E^0 + \frac{RT}{nF} \ln \frac{[\mathrm{Ox}]_0}{[R]_0} \qquad\qquad 3.52$$

When current passes, the concentrations obviously change and a new potential results

$$E = E^0 + \frac{RT}{nF} \ln \frac{[\mathrm{Ox}]}{[R]} \qquad\qquad 3.39$$

subtracting equation 3.39 from 3.52 gives the expression for overpotential.

$$\eta = E - E^e = \frac{RT}{nF} \ln \frac{[\mathrm{Ox}]}{[\mathrm{Ox}]_0} - \frac{RT}{nF} \ln \frac{[R]}{[R]_0} \qquad\qquad 3.53$$

and this is the additional boundary condition for equations 3.44–3.47. It is a boundary condition since it only applies to the surface of the electrode.

Let us now briefly discuss the diffusion overpotential for three cases of experimental interest.

(1) constant current is passing through the cell,

(2) the potential is changed linearly with time, and

(3) the special case of the dropping mercury electrode.

For galvanostatic experiments (those done by passing a constant current through the cell) the relation between potential and time is

$$E = E^0 + \frac{RT}{nF} \ln \left(\frac{D_R}{D_{Ox}}\right)^{\frac{1}{2}} + \frac{RT}{nF} \ln \frac{nF(\pi D_{Ox})^{\frac{1}{2}} [Ox]_0 + 2it^{\frac{1}{2}}}{nF(\pi D_R)^{\frac{1}{2}} [R]_0 - 2it^{\frac{1}{2}}} \qquad 3.54$$

When the constant current is anodic (oxidizing) it is positive and the potential becomes more positive with time. When the current is cathodic (reducing) it is negative and the potential becomes more negative with time. A complete curve for equation 3.54 is given in Fig. 10a. There the following values

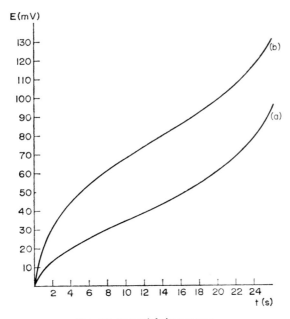

FIG. 10. Potential-time curves.

were used: $E^0 = 0\,V$, $D_R = D_{Ox} = 10^{-5}\,cm^2\,s^{-1}$, $[Ox]_0 = [R]_0 = 10^{-6}$ moles cm^{-3}, $i = 5 \times 10^{-4}\,A\,cm^{-2}$, $n = 1$. The first two terms on the right hand side are equal to $E_{\frac{1}{2}}$ as defined by equation 3.41.

Often, in practical cases, either $[Ox]_0$ or $[R]_0$ are zero. If $[Ox]_0 = 0$ and the electrode process is oxidation, it is convenient to define a "transition time"

$$\tau_R^{\frac{1}{2}} = \frac{nF(\pi D_R)^{\frac{1}{2}} [R]_0}{2i} \qquad\qquad 3.55$$

and equation 3.54 becomes

$$E = E_{\frac{1}{2}} + \frac{RT}{nF} \ln \frac{t^{\frac{1}{2}}}{\tau_R^{\frac{1}{2}} - t^{\frac{1}{2}}} \qquad\qquad 3.56$$

This equation is plotted in Fig. 10b. A similar definition for τ_{Ox} can be given for reduction processes and an equation, equivalent to 3.56, can be derived.

The complete expression for the current voltage curve where the potential is varied linearly with time, is complicated and will not be presented here. However, the important features of this curve will be discussed qualitatively. In order to do this let us consider first that region of potential where limiting current is expected, then the region of low overpotential and finally let us join the two regions. In the region of the limiting current, the concentration at the surface of the electrode is zero, raising the potential will not further change the rate of the reaction because it is diffusion limited at this potential range. The thickness of the diffusion layer, however, increases constantly under these conditions and the current will, therefore, drop as in the case of constant potential, proportional to the square root of time. In the region of low overpotential the current will rise as the potential is raised. If the current

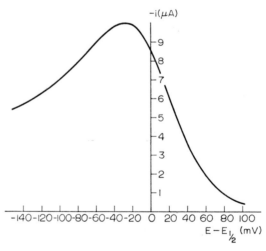

FIG. 11. Current-potential curve for a reduction process on a planar stationary electrode.

rises at low overpotential and falls at high overpotential, the current voltage curve will be peak shaped, as shown in Fig. 11 for a reduction process. The peak current density, i_p is

$$i_p = \pm kn^{\frac{3}{2}} D^{\frac{1}{2}} c_0 b^{\frac{1}{2}} \qquad 3.57$$

where b is the rate of potential change

$$E = E_i + bt \qquad 3.58$$

E_i is the initial potential and b is negative for reduction and positive for oxidation. k is a numerical constant ($2 \cdot 72 \times 10^5$ at 25°C) and D and c_0 refer to the diffusion coefficient and bulk concentration of the reactant. The sign of the current is chosen according to the electrode reaction—positive for oxidation, negative for reduction.

The dropping mercury electrode (d.m.e.) was mentioned in chapter 2B as the best available electrode for the study of the interphase at about room temperature and indeed, the d.m.e. occupies a special place in the study of electrode processes in general. It has four advantages: (1) Mercury is easily purified by distillation and since the drops fall and new ones form, there are new, clean electrodes every few seconds. (2) The drops formed are reproducible. At a given time after the fall of the last drop, the mercury sphere attached to the capillary has the same area each time. The reproducibility of the drops is much better than that of solid electrodes. (3) Overpotential of the reduction of hydrogen is over two volts, making a potential range between zero and $-2 \cdot 5$ volts available for the study of electrode reactions in water. No other metal has such high hydrogen overpotential. However, at approx. zero volts versus the saturated calomel reference electrode, mercury oxidizes to mercurous ions and thus limits the useful potential range there. (4) The d.m.e. is easily constructed and used, every laboratory can have one. This is the reason why it enjoys such a large popularity among workers in electrodics. Since every few seconds a new electrode is formed, the d.m.e. is justifiably treated by potentiostatic methods when the rate of change of potential is not too high. Since the observed current with the d.m.e. varies with the change of area of the electrode, it is better to consider the total current, rather than current density. As a first approximation we may treat the growing sphere (the drop) as a growing plane, and introduce an expression for the expanding area of the electrode into equation 3.48 or 3.49. The area of the electrode is found from the area of the sphere

$$S = 4\pi r(t)^2 = 4\pi \left(\frac{mt}{(\frac{4}{3})\pi\rho} \right)^{\frac{2}{3}} = km^{\frac{2}{3}} t^{\frac{2}{3}} \qquad 3.59$$

where $r(t)$ is the radius of the electrode at any time during its life, m is the rate of flow of the mercury in kilograms per second and ρ is the density in kilograms per metre cube. The radius of the electrode is calculated from the volume, remembering that the rate of flow of mercury in $m^3 s^{-1}$ is m divided by the density. Multiplication of both sides of equation 3.48 or 3.49 by the area yields

$$I = iA = nF(c-c_0)(D/\pi t)^{\frac{1}{2}} k(mt)^{\frac{2}{3}} = k'(c-c_0) D^{\frac{1}{2}} m^{\frac{2}{3}} t^{\frac{1}{6}} \qquad 3.60$$

This is the Ilkovic equation, which was developed by solving the diffusion equations directly very early in the history of polarography. It shows that the diffusion current varies with $t^{\frac{1}{6}}$. (Obviously the maximum current will depend on $t_d^{\frac{1}{6}}$, where t_d is the duration of drop life.) It also shows that the current depends on the rate of flow of mercury. However, the quantity m is not very easily determined, although it can be measured if needed. The parameter that can be measured easily is the height of the surface of the mercury in the reservoir (Fig. 12) above the tip of the capillary, h. Poiseuille's equation states that m is proportional to h and that t_d is inversely proportional to h. Thus $m^{\frac{2}{3}} t_d^{\frac{1}{6}}$ is proportional to $h^{\frac{1}{2}}$. Thus, in order to determine whether a current measured with the d.m.e. is diffusion controlled, all that is necessary is to check whether I is proportional to both $t^{\frac{1}{6}}$ and $h^{\frac{1}{2}}$.

The current actually measured and reported in polarography is not the instantaneous one but an average current

$$I_{\text{average}} = \frac{1}{t_d} \int_0^{t_d} I(t)\,dt \qquad 3.61$$

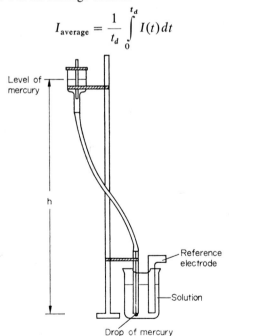

Level of mercury

h

Reference electrode

Solution

Drop of mercury

FIG. 12. The dropping mercury electrode.

It is quite clear from the above that diffusion overpotential gives no information at all on the kinetic parameters of an electrode reaction. Thus if we want to measure kinetics, but find curves which exhibit diffusion characteristics, then the technique must be changed to a faster one, so that the electrode reaction itself would not have time to reach equilibrium. It is the reaction which reaches virtual equilibrium that is mass-transport (diffusion) controlled and where the overpotential gives no information on the kinetic parameters.

D. Overpotential due to slow electron transfer

This section will deal with the "heart" of reactions on electrodes, that of the electron transfer. We shall ask ourselves what is the current–potential relationship when the slowest step in the reaction is the electron transfer step.

Look again at reaction II written as an oxidation

$$cC = aA + ne \qquad\qquad II$$

and, assuming $\bar{\alpha} = \bar{\gamma} = 1$ $\vec{\gamma} = \vec{\alpha} = 0$ (that is: $a = c = 1$), write the rate expression (equation 3.13)

$$i = nF\vec{k}[C] \exp - \bar{\beta}\,\frac{F}{RT}\,\phi - nF\bar{k}[A]\exp - \bar{\beta}\,\frac{F}{RT}\,\phi \qquad 3.62$$

If we write the same reaction in our usual form

$$Red = Ox + ne \qquad\qquad III$$

equation 3.13 will be written as

$$i = nF\vec{k}[R]\exp\frac{\beta nF}{RT}E - nF\bar{k}[Ox]\exp-\frac{\alpha nF}{RT}E \qquad 3.63$$

where E, the electrode potential as measured against a reference electrode, replaces ϕ and therefore

$$\vec{k} = \vec{k}\exp\frac{\beta nF}{RT}\phi_R \qquad 3.63a$$

$$\bar{k} = \bar{k}\exp-\frac{\alpha nF}{RT}\phi_R \qquad 3.63b$$

ϕ_R is the potential of the reference electrode. $-\beta n$ and αn correspond to $\vec{\beta}$ and $\bar{\beta}$, which have been replaced to conform with generally accepted notation.

Equation 3.63 is known as the Butler–Volmer equation. It was derived early in the history of the study of electrode processes following the experimental observations by Tafel (in 1905) that the current density is proportional to the exponent of potential, when E is not too near the equilibrium potential, E^e.

$$i = a \exp{(bE)} \qquad\qquad 3.64$$

Let us now proceed to develop the Butler–Volmer equation and study its implications.

At equilibrium the current flowing through the cell is zero and thus

$$\vec{k}[R] \exp{\frac{\beta nF}{RT}} E^e = \overleftarrow{k}[Ox] \exp{-\frac{\alpha nF}{RT}} E^e \qquad\qquad 3.65$$

Each side of this equation presents the rate of the oxidation or reduction reaction at dynamic equilibrium per unit area of electrode. The current density corresponding to this rate is called the "exchange current density" (i_0).

$$i_0 = nF\vec{k}[R] \exp{\frac{\beta nF}{RT}} E^e = nF\overleftarrow{k}[Ox] \exp{-\frac{\alpha nF}{RT}} E^e \qquad\qquad 3.66$$

Since the Nernst equation is correct at equilibrium, it follows from equations 3.66 and 3.39 that

$$\alpha + \beta = 1$$

$$\frac{RT}{nF} \ln{\frac{\vec{k}}{\overleftarrow{k}}} = E^0$$

Thus dividing equation 3.63 by i_0 yields

$$\frac{i}{i_0} = \exp{\frac{\beta nF}{RT}}\eta - \exp{-\frac{(1-\beta)nF}{RT}}\eta \qquad\qquad 3.67$$

which is the basic equation of electrode kinetics. Thus we have two parameters which characterize the electrode reaction: i_0, the exchange current density and β, which is called the "symmetry factor", "transfer coefficient" or "exponential coefficient". Small differences exist in the exact definitions of these terms, but some authors tend to use them interchangeably. Both these parameters are positive quantities. If the process under consideration is oxidation, η is positive, the first right-hand-side term in equation 3·67 is the greater one and i is positive. If the studied process is reduction, the greater term will be the second and the current is negative.†

---

† This convention (current positive for oxidation and negative for reduction) had not always been the acceptable one. There are many works, particularly early ones, where reduction current is considered positive and vice versa.

The exchange current density obviously depends on concentration. It is desirable to investigate this dependence in order to know what to expect from simple electrode reactions with electron-transfer overpotential. By introducing the value of E^e as given by Nernst equation into both sides of equation 3.66, opening brackets and simplifying in a straightforward way, one gets the following result

$$i_0 = nF\vec{k}[R]^{1-\beta} [Ox]^\beta \exp\frac{\beta nF}{RT} E^0$$

$$= nF\vec{k}[R]^{1-\beta} [Ox]^\beta \exp-\frac{(1-\beta)nF}{RT} E^0 = nFk_s [R]^{1-\beta} [Ox]^\beta \qquad 3.68$$

and k_s is the heterogeneous rate constant of the reaction.

A casual observation of the concentration dependence of i_0 reveals a rather strange feature: the right-hand side of equation 3.68 predicts a "reaction order" of β or $(1-\beta)$; equation 3.66, on the other hand predicts a "reaction order" of one! In order to resolve this seeming paradox, we must remember that i_0 is a function of both [Ox] and [R]. E^e is also a function of both [Ox] and [R]. Thus the partial derivative $\partial i_0/\partial[R]$ or $\partial \ln i_0/\partial \ln [R]$ must specify which quantity, [Ox] or E^e is held constant

$$\left(\frac{\partial \ln i_0}{\partial \ln [R]}\right)_{E^e} = 1 \qquad 3.69$$

from equation 3.66. From equation 3.68.

$$\left(\frac{\partial \ln i_0}{\partial \ln [R]}\right)_{[Ox]} = 1-\beta \qquad 3.70$$

and both are correct. From this discussion it is clear that the term "reaction order" is rather ambiguous when applied to electrode kinetics. If one wants to use this term, one must specify the definition.

In low overpotentials, i.e. small η, one can expand the exponentials of equation 3.67 and neglect all but linear terms:

$$i = i_0 \frac{nF}{RT} \eta \qquad 3.71$$

This affords a very convenient way of measuring i_0, but one cannot get any information about β. To find β, one must measure i_0 for several concentrations of reactant Ox, holding the concentration of Red constant and plot $\log i_0$ against $\log [Ox]$. The slope of this curve is β. If [Ox] is held constant and $[R]$ is varied, the slope of the line $\log i_0$ against $\log [R]$ will be $1 - \beta$.

When the reaction is very slow, so that measurements at low enough over-potential cannot be made and one must use high overpotentials to get measurable current, so high that only one of the directions of the electrode reaction is important, equation 3.67 becomes

$$i = i_0 \exp \beta n F \eta / R T$$

or:

$$i = -i_0 \exp -(1 - \beta) n F \eta / R T \qquad 3.72$$

A plot of $\log |i|$ against $|\eta|$ gives a straight line with slope of $\beta n F / R T$ or $(1 - \beta) n F / R T$ and intercept of i_0. This plot is known as a Tafel plot.

Another way of finding i_0 and β is to write equation 3.67 as:

$$\frac{i}{\exp (n F \eta / R T) - 1} = i_0 \exp \beta n F \eta / R T \qquad 3.73$$

taking logarithms

$$\ln \frac{i}{\exp (n F \eta / R T) - 1} = \ln i_0 + \beta n F \eta / R T \qquad 3.74$$

A plot of the left hand side against η gives a straight line, regardless of whether the overpotential is great or small. The exchange current density is calculated

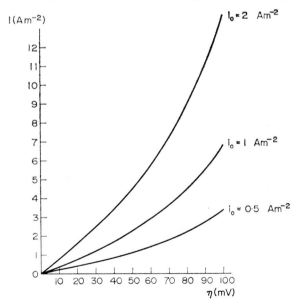

FIG. 13. Current-potential curves from equation 3.67; $n = 1$, $T = 25°C$, $\beta = 0.5$.

from the intercept and β, from the slope. Figures 13 and 14 show calculated current voltage curves with various values of β, and some calculated curves for several values of i_0. Notice the difference between these curves and the ones for "reversible" processes in Fig. 9: while the rise in current at low over-

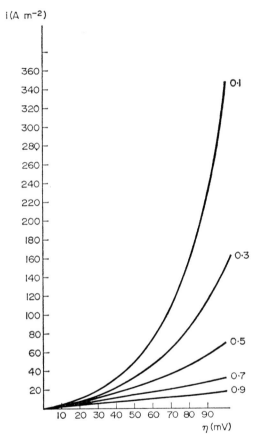

FIG. 14. Current-potential curves from equation 3.67; $n = 1$, $T = 25°C$, $i_0 = 10$ A m^{-2}
The numbers indicate β values.

voltage is very rapid in the "reversible" case and slow in the "irreversible" one, the reverse is true for high overvoltages.

Quite often the electrode reaction is not a simple electron transfer. The electroactive species must often be generated in the solution, the product of electron transfer may undergo further changes before it reaches its final form. When these chemical reactions occur much more rapidly than the electron transfer itself, we may use equilibrium expressions for the preceding or following chemical reactions. Let us deal with the case of

$$qQ + rR \overset{K_1}{=} sS + Ox \qquad\qquad\qquad \text{VIII}$$

$$Ox + ne = Red \qquad \text{rate determining step} \qquad \text{IX}$$

$$Red \overset{K_2}{=} uU + wW \qquad\qquad\qquad X$$

The concentrations $[R]$ and $[Ox]$ for the Butler–Volmer equation are expressed via the equilibrium constants K_1 and K_2.

$$[Ox] = K_1 \frac{[Q]^q [R]^r}{[S]^s} \qquad\qquad [R] = K_2 [U]^u [W]^w \qquad 3.75$$

The exchange current now becomes

$$i_0 = k_s K_1^{1-\beta} [Q]^{q(1-\beta)} [R]^{r(1-\beta)} [S]^{-s(1-\beta)} K_2^{\beta} [U]^{u\beta} [W]^{w\beta} \qquad 3.76$$

q, r, s, u and w are the reaction orders of the reactants and products. They can be determined by measuring i_0 at various concentrations of each of the reactants, holding all other concentrations constant. If we call the concentration of any one of the reactants c_i, we may write it as

$$\left(\frac{\log i_0}{\log c_i} \right)_{c_{j \neq i}} \begin{array}{l} = (1-\beta)\gamma_i \quad \text{For reactants} \\ = \beta\gamma_i \quad\quad \text{For Products} \end{array} \qquad 3.77$$

Where γ_i is the reaction order with respect to reactant i.

From the above treatment can be seen the complexity of the equations for electrode reactions which involve more than one step. When there is more than one electron transfer step, the equations are even more complicated. Equations for reactions which have more than one rate determining step are very involved indeed. Treatment of some of the simpler cases can be found in the literature.

We cannot deal with kinetics without mentioning the activation energy. The definition of A, the activation energy is

$$k = k^0 \exp - \frac{A}{RT} \qquad\qquad 3.78$$

and it looks simple enough, one can measure $\vec{k}$ or $\overleftarrow{k}$ over a range of temperatures, and calculate A in the usual way. However, a look at equations 3.63 (which defines k in measurable terms) and 3.63a, 3.63b (which define k in unknown terms) reveals a problem. If k is measured at a constant potential then, in addition to the temperature dependence of k, there is the temperature dependence of ϕ_R, which is both unknown and dependent on the nature of the reference electrode. Therefore, values of activation energy measured at constant potential over a range of temperatures is meaningless.

The rate constant k_s

$$k_s = \vec{k} \exp \frac{\beta n F}{R T} E^e = \bar{k} \exp - \frac{(1-\beta) n F}{R T} E^e$$

does not depend on the reference electrode and has none of the faults of $\vec{k}$ or $\bar{k}$. Thus an activation energy defined by

$$k_s = k_s{}^0 \exp - \frac{A^{(e)}}{R T} \qquad 3.79$$

is meaningful. But $A^{(e)}$ will be the same for the anodic and cathodic processes for any kinetically reversible reaction. Information may be lost on the individual properties of either process.

If we multiply and divide the right-hand side of equation 3.63 by the respective expressions for the exchange current density

$$i_0 = nF\vec{k}[R] \exp \beta \frac{nF}{RT} E^e = nF\bar{k}[Ox] \exp - (1-\beta) \frac{nF}{RT} E^e$$

we get

$$i = nF\vec{k}[R] \exp \beta \frac{nF}{RT} E^e \exp \beta \frac{nF}{RT} \eta$$

$$- nF\bar{k}[Ox] \exp - (1-\beta) \frac{nF}{RT} E^e \exp - (1-\beta) \frac{nF}{RT} \eta \quad 3.80$$

An analysis of the rate constants of equation 3.80

$$\vec{k}' = \vec{k} \exp \beta \frac{nF}{RT} E^e$$

$$\bar{k}' = \bar{k} \exp - (1-\beta) \frac{nF}{RT} E^e$$

using equations 3.63a and b shows that k' is independent of the reference electrode and, therefore, an activation energy for a constant overpotential

$$\left. \begin{array}{l} \vec{k}' = \vec{k}'^0 \exp - A^{(\eta)}/RT \\ \bar{k}' = \bar{k}'^0 \exp - A^{(\eta)}/RT \end{array} \right\} \qquad 3.81$$

is both measurable and meaningful. $A^{(\eta)}$ will, of course, be a function of the potential.

The above discussion shows very clearly that the idea of activation energy in electrode reactions is not at all the straightforward simple concept of solution kinetics. The difficulty of interpretation, as well as experimental difficulties are to blame for the lack of data on this subject in the electrodics literature.

E. "Kinetic currents", reaction controlled overpotential

Let us now discuss in detail the shape of the current potential curve when the electron transfer step, which is rapid, is preceded by a slow chemical reaction. Consider the following sequence

$$qQ = p \, Red \qquad\qquad XI$$

$$Red = Ox + ne \qquad\qquad XII$$

Since the electron transfer step is rapid, i.e. virtually in equilibrium, the relation between potential and concentrations of the electroactive species is given by Nernst equation

$$\eta = E - E^e = \frac{RT}{nF} \ln \frac{[Ox]}{[Ox]_0} - \frac{RT}{nF} \ln \frac{[R]}{[R]_0} \qquad 3.82$$

Assume, for the sake of simplicity, that the concentration of Q and of Ox are, at the beginning of the experiment, very large compared to that of Red; thus the first term on the right-hand side of equation 3.82 can be omitted. The concentrations of Red can be derived from the rate expression for reaction XI†

$$v = \vec{k}[Q]^q - \overleftarrow{k}[R]^p \qquad 3.83$$

and from the equilibrium constant K

$$K = \frac{[R]_0^{\,p}}{[Q]^q} = \frac{\vec{k}}{\overleftarrow{k}} \qquad 3.84$$

so that

$$\left(\frac{[R]}{[R]_0}\right)^p = \frac{1/\overleftarrow{k}(\vec{k}[Q]^q - v)}{K[Q]^q} = 1 - \frac{v}{\vec{k}[Q]^q} \qquad 3.85$$

and since reaction XI is rate determining,

$$v = \frac{i}{nF} \qquad 3.86$$

† The reaction orders are assumed to be the same as the stechiometric coefficients of reaction XI to keep treatment of the problem as simple as possible.

Combining equations 3.82, 3.85 and 3.86 gives the sought expression for the current potential curve

$$i = \pm nF\vec{k}[Q]^q \left(1 - \exp - \frac{pnF}{RT}\eta\right)$$ 3.87

A plot of equation 3.87 with various values of p, q and $\vec{k}$ is shown in Fig. 15. The values of $\vec{k}$ where chosen so that $nF\vec{k}[Q]^q$ is a constant: $[Q] = 1$ mole m^{-3}, $\vec{k} = 10^{-4}$ ms^{-1} for $q = 1$ and $\vec{k} = 10^{-4}$ m^4 mole^{-1} s^{-1} for $q = 2$. Therefore curves A and B in Fig. 15 approach the same value of limiting current. When $p = 1$ this approach is rather gradual and the value of the limiting current is

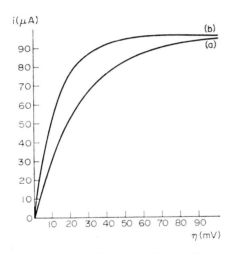

FIG. 15. Current-potential curves from equation 3.87 (reaction overpotential); $[Q] = 1$ mole m^{-3}. Curve (a): $p = q = 1$; $\vec{k} = 10^{-4}$ m s^{-1}; $q = 2$ $p = 1$ $\vec{k} = 10^{-4}$ m^4 s^{-1} mole^{-1}. Curve (b): $p = 2$ $q = 1$ $\vec{k} = 10^{-4}$ m s^{-1}; $p = q = 2$ $\vec{k} = 10^{-4}$ m^4 s^{-1} mole^{-1}.

not attained even when $\eta = 0·1$V. When $p = 2$, the rise in current is much sharper and i attains its essential limiting value at $\eta = 0·065$V. Not shown in the figure, but evident from equation 3.87 is the fact that when η is negative and the electrode process is overall reduction, the effect of the chemical step which follows becomes less pronounced as η increases and we cannot see from the current potential curve that this is not a simple electrode reaction.

The limiting current of reaction sequence XI and XII is less than the corresponding diffusion limiting current had species Q been oxidized directly. This will serve as a good criterion to distinguish between reaction and diffusion controlled electrode processes. This limiting current will be constant even

with solid electrodes in quiescent solutions, contrary to the diffusion controlled case. With the d.m.e. one would observe no dependence of the
current on the height of the mercury reservoir and the current growth at each
drop will be proportional to $t^{\frac{2}{3}}$.   These results are obtained, as before, by
multiplying the two sides of equation 3.87 by the area of the growing
mercury drop.

The above is simple, straightforward and illustrates the characteristic
features of reaction overpotential. However, it has a severe drawback, which
becomes immediately apparent when equation 3.86 is considered more
closely. Consider the dimensions: the reaction velocity is the change in
concentration in time

$$[v] = \frac{\text{moles}}{m^3 . s}$$

and the dimensions of the right-hand side are

$$\left[\frac{i}{nF}\right] = \frac{C}{s . m^2} \frac{\text{mole}}{C} = \frac{\text{moles}}{m^2 . s}$$

The dimensions differ by m.

This discrepancy does not exist when the chemical reaction is heterogeneous, i.e. when Red is strongly adsorbed and $[R]$ is a surface concentration
(moles m^{-2}). But when the reaction is homogeneous there is a clear fault,
the value of $\vec{k}$ measured in this way will not be meaningful. Why is this so?
When a homogeneous reaction near an electrode is considered, the reaction
takes place at a layer near the electrode and the reaction product (in this
case Red) diffuses to the electrode and reacts. In our argument we did not
consider diffusion at all. This was done in order not to mask the essential
points with mathematical detail. When the equations are derived correctly
and the transport of Red to the electrode is considered, one arrives at the
following expression for the current potential curve

$$i' = \pm nF \left(\frac{2}{p+1} \vec{k}[Q]^q D_R[R]_0\right)^{\frac{1}{2}}$$

$$\left\{p + \exp - \frac{nF(p+1)}{RT} \eta - (p+1) \exp \frac{nF}{RT} \eta\right\}^{\frac{1}{2}} \quad 3.88$$

The sign of the current is the same as the sign of the overpotential. When η
is very high and positive, the second term on the right-hand-side becomes
equal to p and the limiting current is

$$i_1' = \pm nF \left(\frac{2p}{p+1} \vec{k}[Q]^q [R]_0 D\right)^{\frac{1}{2}} \quad 3.89$$

The equivalent expression for the limiting current derived from equation 3.87 is

$$i_l = \pm nF\vec{k}[Q]^q \qquad 3.90$$

The dimensions of i_l and i_l' as those of i and i' differ by m.

It is useful to introduce here a concept of the reaction layer thickness (δ_r). As in the case of diffusion, we define it as that distance from the electrode where the concentration of Red differs appreciably from that in the bulk.

$$i_l' = i_l \delta_r, \qquad i' = i\delta_r \qquad 3.91$$

On introducing equations 3.89 and 3.90 into 3.91, the value of the reaction layer thickness for large overpotentials is calculated to be

$$\delta_r = \left\{ \frac{2p[R]_0 \, D}{(p+1)\,\vec{k}[Q]^q} \right\}^{\frac{1}{2}} \qquad 3.92$$

For small overpotentials the exponents can be expanded into Tailor series. The result for low overpotential is

$$\delta_r = \left\{ \frac{[R]_0 \, D}{p\vec{k}[Q]^q} \right\}^{\frac{1}{2}} \qquad 3.93$$

It is seen from the above that δ_r depends on the overpotential, i.e. on the current density, except when $p = 1$.

F. The influence of surface structure on electrode processes

The electrode reactions considered so far in this chapter did not involve adsorbed reactants or products. However, it is unrealistic to expect that most actual processes are so simple. Therefore let us deal now with the effect of various types of adsorption on electrode processes. First let us classify adsorption phenomena.

The most common classification is between physical and chemical adsorption. Physical adsorption is caused by the tendency of all gases, in temperatures below their critical temperatures to condense to some extent. Gas will condense preferentially on a surface rather than in the form of droplets because the vapour pressure of small drops is higher than that of a flat layer of liquid. Thus, the standard free energy of physical adsorption is of the same magnitude as that for the liquification of gases (2–5 kcal mole^{-1}). In chemical adsorption, on the other hand, there is formation of bonds between the substrate (in our case, the electrode) and the adsorbate. Thus, the standard free energy of adsorption ΔG_{ads}^0 is of the same magnitude as that for chemical reactions. Hence, chemical adsorption cannot form more than a monolayer of the species on the electrode, while physical adsorption is not restricted in this way. It is convenient to think of the energy of adsorption on electrodes as composed of two parts, a "specific" part, which is independent of potential, and the "non specific" part, which is potential dependent

$$\Delta G^0_{ads} = \Delta G^0_{sp} + \Delta G^0_{nsp}(E) \qquad\qquad 3.94$$

$\Delta G^0_{nsp}(E)$ depends on the charge of the molecule and on its structure. The exact dependence is not generally known, but useful functions for ions are

$$\Delta G^0_{nsp}(E) = \pm g(E - E_{pzc}) \qquad\qquad 3.95$$

where g is a constant of proportion and includes the charge of the ion, the concentration of solution and the natures of solvent and solute. E_{pzc} is the potential of zero charge of the electrode, and the sign is $+$ for cations and $-$ for anions. (E_{pzc} is what its name implies—that potential where the charge on the electrode is zero. It will be dealt with in detail in Chapter 4.) The equivalent useful function for neutral molecules is

$$\Delta G^0_{nsp}(E) = \Delta G^0_{max} + g'(E - E_{max})^2 \qquad\qquad 3.96$$

E_{max} is close to the potential of zero charge, but not identical with it. ΔG^0_{max} depends on the solubility of the adsorbate in the cell solution: the less soluble the adsorbate, the more negative is ΔG^0_{max}. This subject of potential dependent adsorption will be discussed again in the next chapter.

A reactant solution in an electrochemical cell usually contains four kinds of species: the reactant, the product, the solvent and the electro-inactive material, often an electrolyte. Each one of these can be adsorbed on the electrode. The adsorption of the solvent is often neglected since the coverage is very close to maximum, $\theta = 1$, and does not change when the cell parameters change.

Let us say a few words here on the influence of the adsorption of the inactive material (or supporting electrolyte) on the kinetics of the electrode reaction. Inactive molecules in an electric field could very roughly be approximated by uncharged conductors in this electric field. Figure 16 shows how the lines of force of a homogeneous electric field are altered by the presence of a metallic conductor. Thus one might expect that an inactive molecule in the field of the double-layer would also change the lines of force there, thereby altering the field strength and the quantities, such as concentrations of ions, that depend on it. Electro-inactive molecules may not be inert but may react with the electroactive molecule (e.g. Red).

$$A + Red = ARed \qquad\qquad XIII$$

where A represents the electro-inactive molecule. If the concentration of A at the surface differs from that in the bulk, perhaps due to adsorption, the concentration of ARed at the surface will also differ from that in the bulk, the result being either a speeding up or a slowing down of the electrode

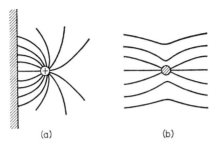

(a)   (b)

FIG. 16. Lines of force in an electric field: (a) A charge near a conducting plane; (b) A conducting sphere in a uniform electric field.

reaction.   We see that adsorption of apparently inactive molecules may influence both the field in the interphase and the concentration of the electroactive species. This subject will be taken up again in Chapter 4 when the influence of the structure of the interphase on electrode kinetics will be discussed.

It remains to consider in this section the influence of adsorption of the reactant or product on the kinetics of the electrode process.

When $\Delta G_{sp}^{0} \gg \Delta G_{nsp}^{0}$, the adsorption may be regarded as potential independent. In this case one can substitute the surface concentration, as obtained from the adsorption isotherm, in the Butler–Volmer equation for the reaction

$$Red(sol) = Red (ads) \qquad\qquad XIV$$

$$Red(ads) = Ox (sol) + ne \qquad\qquad XV$$

Two situations arise here: (1) when reaction XIV is virtually in equilibrium and (2) when reaction XV is the much faster one. The second case is analogous to the kinetic current case in that the kinetics of adsorption are measured rather than that of electron transfer. The relevant equation for the current potential curve has been given (equation 3.87). The first case will be treated here.

Let us write the current-potential equation for reactions XIV and XV; the anodic current will depend on the surface concentration of Red, L_{Red}.

$$i_a = \vec{k} L_{Red} \exp \beta nFE/RT \qquad\qquad 3.97$$

and the cathodic current will, of course, depend on the concentration of Ox in the solution. However, the area of the electrode for the cathodic reaction is not the area of the whole electrode, it is partly blocked by the adsorbed molecules of Red. Since the fraction of area blocked is θ, the fraction available for reduction is $1-\theta$. Therefore this term must be inserted in the expression for the current.

$$i_c = -\vec{k}[\text{Ox}](1-\theta) \exp - (1-\beta) nFE/RT \qquad 3.98$$

Before writing the expression for the net current, $i_c + i_A$, multiply and divide i_A by $L_{max} \exp (\beta nf E^e RT)$, and multiply and divide i_C by $\exp - [(1-\beta)nFE^e/RT]$

$$i = \vec{k}' \theta \exp \beta nF\eta/RT - \overleftarrow{k}'[\text{Ox}](1-\theta) \exp - (1-\beta) nF\eta/RT \qquad 3.99$$

where
$$\vec{k}' = \vec{k}L_{max} \exp \beta nFE^e/RT$$

$$\overleftarrow{k}' = \overleftarrow{k} \exp - (1-\beta) nFE^e/RT$$

It is apparent that when the adsorption is not potential dependent, the shape of the current potential curve will remain the same, and i_0 and β could be determined in the usual way, i.e. by using Tafel plots. However, the expected proportion between i_0 and concentration, at constant equilibrium potential, will not be followed. Instead, i_0 will depend on the bulk concentration of Red in a way characteristic of the adsorption isotherm.

When the adsorption must be regarded as potential dependent (as many cases are) the expression for the potential dependence of ΔG^0_{ads} must be considered. If we consider neutral species which follow the approximate Temkin adsorption isotherm, we get by combining equations 3.94 and 3.96

$$\Delta G^0_{ads} = \Delta G^0_{sp} + \Delta G^0_{max} + g'(E - E_{max})^2 = \Delta G^0 + g'(E - E_{max})^2 \qquad 3.100$$

using equations 3.99 and 3.34 with 3.100 yields

$$i = \frac{\vec{k}'}{b} \{\Delta G^0 + g'(\eta - \eta_{max})^2 + 2 \cdot 3RT \log [R]\} \exp \beta nF\eta/RT$$

$$- \frac{\overleftarrow{k}'}{b} [\text{Ox}]\{b - \Delta G^0 + g'(\eta - \eta_{max})^2 + 2 \cdot 3 \log [R]\}$$

$$\times \exp - (1-\beta) nF\eta/RT \qquad 3.101$$

Quite obviously, a Tafel plot will not give a straight line and is therefore useless. In spite of the complicated form of equation 3.101, several points can be made about it. The exchange current, measured for various concentrations of Red and Ox, so that the equilibrium potential remains constant, should be linear with log [R] and proportional to [Ox]. If E_{max}, b and ΔG^0 are known from independent measurements, $\vec{k}'$ and $\overleftarrow{k}'$ can be evaluated and, hence, also β.

It is thus quite clear that when the electrode reaction is complicated by the adsorption of either the reactant or the product, the kinetic equations become very involved. How difficult is the treatment of these cases will become clearer in the next chapter when we shall consider the influence of adsorbed species on the structure of the double layer, and the influence of the double layer structure on the kinetics of electron transfer.

We see that, unlike the former reactions which dealt with processes whose rate limited the rate of the overall electrode reaction, this section did not deal with processes limited by the rate of adsorption. We studied here the influence of adsorption on reactions where kinetics are given by the Butler–Volmer equation. We saw that if the reactant or the product is adsorbed on the electrode, and this adsorption depends on the potential—then the expected behaviour of a simple electron-transfer overpotential is not observed. There is another class of electrode reaction which does not behave simply, because their rate depends on the availability of suitable sites on the electrode surface. These are the electrocrystallization reactions, in particular, metal deposition reactions.

$$M^+ + e = M^0 \qquad \qquad \text{XVI}$$

M^+ represents a metal cation: M^0—a metal atom. It is well known that the metal atoms, when they reach the electrode surface, do not deposit randomly on the metal solid electrode—the crystal grows preferentially at definite growth sites. Thus, the current will not depend on the concentration of M^+ alone, but also on the concentration of the growth sites on the electrode L_m.

$$\frac{i}{nF} = -\vec{k}[M^+]L_m \exp-(1-\beta)\frac{nF}{RT}E + \overleftarrow{k}L_m \exp \beta \frac{nF}{RT}E. \quad 3.102$$

Let us look at the growth sites more closely and consider factors which affect their concentration on the metal electrode: the composition of the solution and the potential. Like growth sites of any other crystal, those on electrodes are found at the grain boundaries and at crystal dislocations and imperfections. These depend on the history of the electrode, whether it was stretched, hammered or annealed and thus L_m will also depend on the history of the electrode. If the solution contains species that adsorb on the electrode at the crystal growth sites, L_m will be smaller than if the solution was free from such species. This alone demands the exercise of great care in handling experiments and experimental data on electrocrystallization.

We shall now discuss the dependence of L_m on the electrode potential. We know that when the overpotential is small the "attraction" between the ions and the electrode is small, a given ion may be reduced or not and the low reaction rate (at low overpotential) will not be affected too much. When the reaction rate is increased by increasing the overpotential, a stage is reached when all ions at the surface must reduce in order to maintain this high reaction rate. Thus at high overpotentials, all kinks, dislocations and imperfections will be used as growth sites. At low overpotentials only the most prominent imperfections (such as those in Fig. 17c) will be used as growth

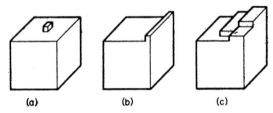

(a)　　　　　(b)　　　　　(c)

FIG. 17. Various sizes of surface imperfections: (a) small, (b) medium, (c) large.

sites. We see that, although the number of imperfections does not change as a function of potential, the number of those imperfections which serve as growth sites does change with potential. Theoretical calculations show that the current for metal deposition at an overpotential η, is proportional to $\exp - a\eta$, $\exp - a/\eta$ or $\exp - a/\eta^2$ depending on the exact mechanism; a contains factors such as the atomic weight of the metal, the edge free energy, the density of dislocations intercepting the surface and the like. Thus a current-potential curve for metal deposition will not necessarily look like a simple Butler–Volmer curve.

Up to now we have been concerned with current-potential curves, but when measuring these we also measure the potential-resistance or impedance curves. What is the cell impedance and how does it behave? This is the subject of the next section.

G. The cell impedance

An impedance Z of any component in an electrical circuit is defined as the ratio of voltage across this component to the current passed through it.

$$Z = E/I \qquad 3.103$$

The impedance is the "resistance" of that component towards alternating current. It is composed of the resistance R of the component, and its reactance; the reactance is again divided to capacitive reactance $(1/\omega C)$ and inductive reactance (ωL). ω is the frequency of the alternating current. In electrochemical cells the inductive reactance is nil and the impedance is divided into resistance and capacitance. The impedance of an electrochemical cell is composed from the impedance of the reference electrode (which is negligible according to the definition of reference electrodes, Chapter 2, D), the resistance of the solution and the impedance of the interphase between the working electrode and the solution. It is this last impedance which is of interest in electrode kinetics. Figure 18 shows the equivalent circuit for an

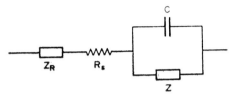

FIG. 18. The equivalent circuit of a cell.

electrochemical cell: the impedance of the reference electrode, Z_R, is usually neglected; the resistance of the solution is R_s; the capacity of the double layer is C and the so called "faradaic impedance" is Z. Z arises from the electrode reaction and we do not know at this stage whether it is pure resistance or includes a reactive component as well. C and Z must be connected in parallel because they offer alternative routes for the flow of current, alternating current flows mainly through C, direct current flows mainly through Z.

Since current passed through a cell is not proportional to the applied voltage, current voltage curves usually are not straight lines, the cell impedance is not constant. As it is more convenient to deal with values of impedance rather than functions, one defines the "Faradaic impedance" not simply as the ratio of voltage to current, but as the limit

$$Z = \lim_{\eta \to 0} \frac{dE}{dI} \qquad 3.104$$

We will now derive the faradaic impedance due to a reaction whose rate is limited by electron transfer.

$$\eta = E - E^e$$

$$d\eta = dE$$

$$\frac{1}{Z} = \lim_{\eta \to 0} \frac{dI}{dE} = \lim_{\eta \to 0} \frac{di}{dE} \frac{1}{A}$$

for constant area electrodes. From the Butler–Volmer equation

$$\frac{di}{dE} = i_0 \left\{ \beta \frac{nF}{RT} \exp \beta \frac{nF}{RT} \eta + (1 - \beta) \frac{nF}{RT} \exp - (1 - \beta) \frac{nF}{RT} \eta \right\} \quad 3.105$$

Taking the limit yields

$$\frac{1}{Z} = i_0 \frac{nF}{RT} \qquad\qquad 3.106$$

Since Z does not include any time dependence it must be a pure resistance. (If an expression for impedance includes time, it can be expected that this impedance will depend on the frequency and, therefore, will not be pure resistance.)

The electrode reaction rate may be controlled by diffusion or by chemical reaction. The impedances associated with these cannot be derived so simply. All that needs to be said here is that they are not pure resistances, i.e. they include a reactive part which, for convenience, is called "pseudocapacity" to distinguish it from the "true capacity" of the double layer. The diffusion impedance is also often referred to as "Warburg impedance".

Let us now turn to the discussion of one of the persistent complications of the experimental study of electrodics: the continuous presence of the solution resistance. The electrolyte is there in studies of direct or alternating currents, it is there at high or low overpotentials, and a way must be found to either calculate or eliminate its resistance. Consider again Fig. 18: when a voltage V is applied across the cell and a current I passes, the voltage is divided between the solution and the electrode surface

$$V = IR_s + IZ \qquad\qquad 3.107$$

(Z being the total impedance of the interphase). If R_s is of the same order of

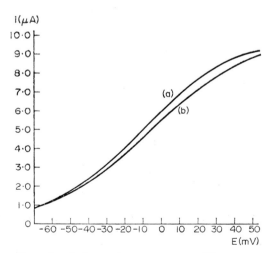

FIG. 19. The effect of solution resistance on reversible current-potential curves: a) $R = 0$, (b) $R = 1000\Omega$

magnitude as Z a fair proportion of the voltage drop will occur across the solution, and only part of it will influence the electrode interphase. Figure 19 shows what happens to a current voltage curve when the solution resistance is 1000 ohms. Curve (a) is calculated from equation 3.43 using $E_{\frac{1}{2}} = 0.0$ volts, $n = 1$ and $I_{la} = 10^{-5}$ amperes, and curve (b) was calculated for the same conditions with solution resistance. It is evident that curve (b) does not represent the correct current voltage curve because of solution resistance and this complication may be serious at times.

There are several ways of minimizing the distorting effect of the solution resistance. The simplest is to measure R and find the potential drop IR by reading I from the current potential curve. The correct potential can now be calculated by using equation 3.107 and a correct current potential curve drawn. This method is, however, very tedious. Another way is to introduce into the cell a reference electrode in addition to the two electrodes in the cell and measure the potential between the working and reference electrodes. Since the current drawn by the potentiometer is extremely small, the solution resistance has no effect on this measurement and the recorded potential is the true one. The problem with this method, however, is that the potential drop in the solution creates a field there and when two con- ductors such as electrodes are introduced into a field a potential difference will always be measured between them. The only way to overcome this is to place the reference electrode as close as practicable to the working electrode. This is usually done by making the tip of the reference half-cell a fine capillary, called "Luggin capillary", and arranging it as close as possible to the working electrode.

There is a further property of the potential drop IR which lends itself to experimental use. When a step current is applied to the cell (i.e. current that increases instantaneously, stays at a constant value for a while and then decreases instantaneously to zero) and the voltage is measured on an oscilloscope, it can be seen that the potential rises to a certain value abruptly at first and then more gradually. This sudden rise is the IR potential. Remember that a pure resistance has no time lag and assumes the proper potential as soon as current is applied, while attainment of steady potential on the electrode surface proper is always slower.

H. Complex electrode reactions; showing the way

Up to now we have discussed only very simple electrode reactions with one well-defined electron transfer step. Real electrode reactions are usually much more complex and may involve several steps of similar rate, parallel reactions and other complications. Let us now look at a general electrode reaction where there are several reactants and products and more than one electron transfer step. However, we still assume that there is only one rate determining step (which is an electron-transfer one) and that there are no parallel reactions.

$$v_1\, S_1 + v_2\, S_2 + \ldots = v_p\, S_p + v_q\, S_q + \ldots + ne \qquad \text{XVII}$$

where v_i are the stechiometric coefficients of species S_i. If we regard the stechiometric coefficients of the reactants as negative, this can be written in shorthand notation as

$$\sum_i v_i\, S_i + ne = 0 \qquad \text{XVIII}$$

Let us assume that there is only one rate-determining step in this reaction but several electron transfer steps. The rate-determining step can be written as

$$(a/v)\, A \rightleftarrows (b/v)\, B + (z/v)\, e \qquad \text{XIX}$$

The reactions preceding this step can be summarized as

$$\sum_i \vec{v}_i\, S_i + aA + \vec{z}e = 0 \qquad \text{XX}$$

and the reactions following the rate-determining step can be summarized as

$$\sum_i \grave{v}_i\, S_i - bB + \grave{z}e = 0 \qquad \text{XXI}$$

it follows from stechiometry that

$$\vec{v}_i + \grave{v}_i = v_i \qquad 3.108$$

$$\vec{z} + \grave{z} + z = n \qquad 3.109$$

There are two basic ways by which complex reactions are treated: (a) by assuming a steady state concentration of the reactant A or (b) by assuming that reactions XX and XXI are virtually at equilibrium and can be treated with equilibrium expressions, i.e. the Nernst equation. Let us proceed using assumption (b). For reaction XX

$$E = \vec{E}^0 + \frac{RT}{\vec{z}F}\, \ln \prod_i [S_i]^{\vec{v}_i}\, [A]^a \qquad 3.110$$

and for reaction XXI

$$E = \grave{E}^0 + \frac{RT}{\grave{z}F}\, \ln \prod_i [S_i]^{\grave{v}_i}\, [B]^{-b} \qquad 3.111$$

E is the electrode potential and $\vec{E}^0$ and $\grave{E}^0$ are standard electrode potentials for reactions XX and XXI, respectively. The current density is given by the Butler–Volmer equation for reaction XIX

$$i = nF/v \left\{ \vec{k}[A]^{a/v} \exp \frac{zF\beta}{vRT}\, E - \grave{k}[B]^{b/v} \exp - \frac{zF(1-\beta)}{RT}E \right\} \qquad 3.112$$

The concentration terms for A and B can be derived from equations 3.110 and 3.111. Substituting these into equation 3.112 yields

$$i = nF/v \left\{ \vec{k} \prod_i [S_i]^{-\bar{v}_i/v} \exp - \frac{\vec{z}F}{vRT} E^0 \exp \frac{\vec{z}+z\beta}{v} \frac{F}{RT} E \right.$$

$$\left. - \overleftarrow{k} \prod_i [S_i]^{\bar{v}_i/v} \exp \frac{\overleftarrow{z}F}{vRT} E^0 \exp - \frac{\overleftarrow{z}+z(1-\beta)}{v} \frac{F}{RT} E \right. \qquad 3.113$$

Thus, the experimental transfer coefficients are

$$\alpha_{anodic} = \frac{\vec{z}+z\beta}{v} \qquad\qquad 3.114$$

$$\alpha_{cathodic} = \frac{\overleftarrow{z}+z(1-\beta)}{v} \qquad\qquad 3.115$$

v is the number of times the rate determining step occurs in one unit electrode reaction, for example $v = 2$ for the hydrogen evolution reaction if the following mechanism operates

$$H_3O^+ + e \rightarrow H(ads) + H_2O \qquad\qquad XXII$$

$$2H(ads) \rightarrow H_2 \qquad\qquad XXIII$$

reaction XXII occurs twice for every unit of the electrode reaction

$$2H_3O^+ + 2e \rightarrow H_2 + 2H_2O \qquad\qquad XXIV$$

It is obvious that for complex electrode processes the anodic and cathodic transfer coefficients do not add up to unity. The determination of the various coefficients, $\vec{z}$, $\overleftarrow{z}$, z and v is usually not straightforward and involves some guesses about the plausible behaviour of the electrode reaction under study.

We now recognize the complexity of electrode reaction kinetics and can ask ourselves how, when studying a particular electrode reaction, we should go about finding the right values for the kinetic parameters i_0 and α, and how they can be analysed in terms of a stepwise mechanism. This section will attempt to give a short guide on the solving of electrodic problems. It must always be remembered, however, that only a very rough guide can be given without knowing the particular electrode reaction studied.

Starting with voltammetry, we get a well defined current–potential curve for the reaction to be studied. The first step is then to find the identity of the product(s). Large scale electrolysis, followed by conventional analytical techniques, often gives the answer. Quantitative coulometry, where we react a known amount of material and measure the electricity consumed, gives the value for n. Often the current voltage curve will exhibit a limiting current region and it is important to know which step will limit the current there. This could be either diffusion or a chemical reaction. This stage is quite easy, when using a solid working electrode, the solution may be stirred. The diffusion current will be affected to a large extent but a reaction controlled current will be only slightly affected, if at all. If slow stirring increases the limiting current, but increased stirring speed has no larger effect, it is clear that there is a preceding reaction whose rate is comparable with diffusion, but not with convective mass transport. If we use the d.m.e. we can use the dependance of the limiting current on drop time and on the height of the mercury column to distinguish between diffusion and reaction controlled limiting current.

If the limiting current is controlled by diffusion, we must see whether the rising portion of the wave is given by the Nernst equation (i.e. the wave is reversible) or by the Butler–Volmer equation. In the first case it is clear that polarography is not suitable for the study of this reaction and other methods (to be described in Chapter 5) must be used. In the second case the exchange current and the transfer coefficient are determined using the appropriate graphical methods (Tafel plot and the like). The study of the concentration dependence of i_0 on the reactant as well as on all other constituents will reveal any chemical reaction at equilibrium that might precede or follow the electron transfer step. It is here that we may find adsorption effects, catalytic or inhibiting effects, or the possibility of two electron transfer steps in the studied electrode reaction. If the transfer coefficients are far from $0 \cdot 5$ one suspects that the reaction is complex. Here starts the guess-work. We may postulate various mechanisms for the reaction at hand attempting to find the correct numbers for the electrode reaction order, the symmetry factor (which is often $0 \cdot 5$) the number of electrons per step and the stechiometric coefficient of the reactant, in order to account for the experimental value of the transfer coefficient.

If the limiting current is found to be reaction controlled, the kinetics of this preceding reaction may be obtained by finding the dependence of this limiting current on the concentrations of reactant, product and other components of the solution. If the rising portion of the wave is described by the Nernst equation, then other methods must be used in order to study the kinetics of the electron transfer step. If the rising portion of the wave is described by the Butler–Volmer equation, we can study the electron transfer step using polarography or voltammetry. The same methods used in the study of the former group of reactions apply here as well.

When conventional polarography is too slow for studying the kinetics of a reaction, we usually turn to methods that give the desired information (i.e. i_0 and α) faster. Very fast recording of current potential curves is not suitable because, when the potential changes very rapidly, much of the current will be capacity current, i.e. that used to charge the double layer. The study of current-time and potential-time curves are better methods as measurements can be made in a fraction of a second, although the actual study may take quite a long time.

The problem with rapid electrode reactions is that they become mass transport controlled at a rather low overpotential, masking any possibility for the determination of kinetic parameters. In such cases it is desirable to have methods which require only very small overpotential so that diffusion is still a minor factor in the kinetics. The measurement of cell impedance at the equilibrium potential gives, under certain conditions, such a possibility.

In this chapter we have tried to give the reader some understanding of electrode kinetics and the way in which the various expressions for reaction rates are derived. Electrode kinetics being such a large and complicated field, it is recommended that this information be supplemented by more advanced books if the reader is interested in working in this area. However, let us summarize by the following points

(i) There are four basic processes that should be considered: mass transport, electron transfer, accompanying chemical reactions and adsorption or crystallization.

(ii) Every kind of mechanism should be dealt with individually and appropriate equations derived in view of the available experimental data.

(iii) The technique used should match the reaction studied as to speed and sensitivity.

4. The Electrode-Solution Interphase at Equilibrium

A brief introduction to the structure of the interphase was given in Chapter 2, Section B. This chapter will be devoted to a detailed study of the interphase, its properties and the ways in which it influences electrode kinetics. We shall restrict ourselves to liquid electrodes, i.e. mercury, since it is with the d.m.e. that most experimental work on the interphase has been done.

A. A qualitative presentation of the interphase

First let us turn our attention to the metal surface and understand the simplifications which we usually make when considering it as part of the electrode-solution interphase; then we shall present a model for the structure of the solution at the interphase and discuss the expected behaviour of this model when the potential of the electrode varies. We shall continue to discuss the influence that the interphase may have on electrode reactions and conclude this section by enumerating the different variables which are important in the study of the interphase.

The metal part of the interphase probably consists of several layers of ions embedded in the electron "cloud" of the metal. However, the electrode surface is always presented as a continuous plane and the charge on it is thought of as being uniformly distributed. This presentation is obviously not complete, the ions of the metal are of similar size to the molecules of the solution and the surface should strictly be presented as a net rather than a plane. Moreover, the charge is not uniformly distributed, it comes in "packets" of electrons and cations whose charge is of the same order of magnitude as the charge on the solution particles. Ideally, we should consider the structure of the metal electrode and the discrete nature of the charge carrying species. However, the sacrifice of accuracy of the model results in simple straightforward derivations of expressions for potential and concentration in the interphase which give fair agreement between theory and experiment in simple cases. In this discussion, the metal electrode will be presented as a continuous plane in which the charge is uniformly distributed.

Figure 20 represents the structure of the solution side of the interphase as it is understood today. Four types of solution species are shown, cations (shown with their solvation sheath), anions, solvent molecules (shown as small dipoles) and large, neutral molecules. The model shows that the anions are not solvated; this is in agreement with experimental results on the

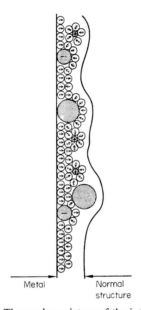

FIG. 20. The modern picture of the interphase.

hydration numbers of cations and anions: cations are always much more strongly hydrated than anions of similar size. Thus, anions approach the electrode surface to a smaller distance than cations. The greater fraction of the electrode is covered by solvent molecules because these are so abundant in the solution. These are shown in Fig. 20 with all the dipoles oriented in the same direction. In reality some will be oriented in the reverse direction, but one can expect that the layer of dipoles closest to the electrode will be largely uniformly oriented because of the existence of surface forces acting on this layer (see Chapter 2, Section B).

When the electrode is charged, the concentration of ions, as well as the orientation of dipoles at the interphase will be affected in a way calculable from electrostatistics (see Section B). This influence of the electrode charge on concentration of ions and on the orientation of dipoles depends on the charge of the ion, on the dipole moment of the molecule and on the magnitude of the electric field present.

Some of the species in solution will adsorb on the electrode. Let us now define adsorption and divide it into what is known as the "specific" and "coulombic" parts. A species is considered adsorbed when its concentration in the interphase region differs from that in the bulk. It is called "positive" adsorption if the concentration in the interphase is higher than in the bulk, or "negative" adsorption if the concentration is smaller. One factor which influences adsorption is the charge on the electrode—if the charge is positive, anions will be attracted to it and cations will be repelled from it. This part of adsorption is the "coulombic" adsorption and can be calculated from appropriate models of the interphase. Any positive adsorption that cannot be accounted for by coulombic forces is called "specific" adsorption. Specific adsorption results when the energy of interaction between the adsorbate and solvent is (a) smaller than the energy of interaction of adsorbate molecules with themselves or (b) when it is smaller than the energy of interaction between the adsorbate and the electrode material. Thus, it can be expected that species which are not very soluble in the solvent will adsorb on the electrode in the same way as they adsorb in the solvent-air interphase. It can also be expected that molecules which interact with the conduction band of the metal will adsorb in the metal-solution interphase and not at the solution-air interphase. The forces which cause specific adsorption are called "specific forces".

It is evident from Fig. 20 that the distance of closest approach, i.e. the distance between the electrode surface and the centre of the molecule adjacent to it, varies with the type of the molecule (or ion), e.g. the distance of closest approach is large for large species and small for small species. It seems likely, however, that specifically adsorbed species will be closer to the electrode than non-specifically adsorbed ones. Hence one may expect a layer of solvent molecules between non-specifically adsorbed species and the electrode. It is a convenient simplification to imagine a plane which passes through the centres of the specifically adsorbed species (excluding the solvent) and call that the "inner Helmholtz plane" (i.h.p.). In a similar way we can imagine a plane passing through the centres of the non-specifically adsorbed species and call that the "outer Helmholtz plane" (o.h.p.). The o.h.p. is usually considered to be the site of the electrode reaction, although, for specifically adsorbed species, the site of the reaction is the i.h.p. When we think of the inner electrode potential ϕ, we think of the work needed to bring a unit charge from the bulk of solution to the inside of the electrode. In a similar way we may think of the potentials at the i.h.p. and the o.h.p., $\phi_{i.h.p.}$ and $\phi_{o.h.p.}$, as the work needed to bring unit charges from the bulk of the solution to the i.h.p. and to the o.h.p. respectively. $\phi_{i.h.p.}$ and $\phi_{o.h.p.}$ will differ in general from the electrode potential ϕ.

Consider now a cell made up of the hydrogen standard electrode and the electrode under discussion. As the electrode potential, E, changes from positive to negative values, the charge on the electrode will also change from positive to negative values. There will be a point where the potential of this cell corresponds to the charge on the electrode being zero. This potential is called "the potential of zero charge" (p.z.c.). It might be expected that at the p.z.c., the potential across the interphase is zero and, consequently, the electrode potential ϕ (which is the potential difference between the electrode and the bulk of the solution) could be determined. However, even though the electrode has no charge there is still a potential difference across the interphase arising from the presence of the layer of oriented solvent dipoles and other adsorbed species in contact with the electrode. We may generalize and say that across any interphase there is a layer of oriented dipoles which gives rise to a surface potential χ. Thus the potential across the interphase is composed of two components: one caused by the charge on the electrode and called the "outer potential" ψ and the other, the surface potential. The outer potential can, in principle, be measured or calculated, but the surface potential cannot. Thus, the "inner potential" ϕ cannot be measured since it is given by

$$\phi = \chi + \psi \qquad 4.1$$

The outer potential is a simple function of the charge on the electrode and the charges in the solution and will be calculated in Section B of this Chapter. The surface potential is a complex function of the electrode charge and of the nature and concentration of the solution.

The potentials of zero charge depend on the metal of the electrode, as well as on the composition of the solution. Table II gives the potential of zero

TABLE II. *Potentials of zero charge of metals*

Metal	Solution	E_{pzc} V versus s.c.e.
Platinum	$0.003M$ $HClO_4$	0.17
Gold	$0.01M$ Na_2SO_4	-0.01
Silver	$0.1M$ KNO_3	-0.19
Copper	$0.01M$ Na_2SO_4	-0.21
Antimony	$0.10M$ HCl	-0.43
Mercury	$0.01M$ NaF	-0.43
Cobalt	$0.01M$ Na_2SO_4	-0.56
Bismuth	$0.01M$ KCl	-0.60
Iron	$0.0005M$ H_2SO_4	-0.61
Aluminium	$0.01M$ KCl	-0.76
Lead	$0.01M$ KCl	-0.93
Cadmium	$0.01M$ KCl	-1.16

charge E_{pzc} for various metals and Table III gives the E_{pzc} for mercury in various electrolyte solutions.

TABLE III. *Potentials of zero charge on Mercury in various electrolyte solutions*

Electrolyte	Concentration (M)	$-E_{pzc}$ V versus s.c.e.
KF	0·1	0·43
LiCl	1·0	0·52
NaCl	1·0	0·52
KCl	1·0	0·52
RbCl	1·0	0·52
CsCl	1·0	0·52
KOH	0·1	0·44
KBr	0·1	0·53
KCNS	0·1	0·59
KI	0·1	0·69

It is clear from the Tables that the nature of the electrode influences the potential of zero charge much more than the composition of the solution. Thus the p.z.c. is an intrinsic function of the electrode and can serve as an internal reference of potential for the system. Indeed, in many cases, it is convenient to express the electrode potential as a "rational potential" $E - E_{pzc}$, for example when comparing the interphase behaviour of different metals or when comparing the mechanisms of electrode reactions on several metals, the rational potential gives a better indication of the field strength and concentration of non-specifically adsorbed ions in the interphase than the usual electrode potential E.

There are four ways by which the interphase influences the rate and mechanism of electrode reactions.

(a) As was shown before, molecules and ions cannot approach nearer to the electrode than a small distance, they are of finite size and thus their centres cannot be closer to the electrode than their radii. The site of the reaction is, therefore, that distance away from the electrode. Thus, the potential at the site of reaction differs from the electrode potential. Consider now the way in which equation 3.8 was derived: the "concentration" of electrons was given in terms of the potential and concentration of electrons in a standard state and the electrode potential, ϕ. However, when the electron does not have to go from the electrode to the bulk of the solution but to a point near the electrode, where the reactant is situated and the potential is not that of the bulk of the solution, the energy of this electron will be different from $nF\phi$. If the site of the reaction is the outer Helmholtz plane, the energy of these electrons (per mole of reaction) is $nF(\phi - \phi_{o.h.p.})$, and this expression should be substituted in equation 3.8 instead of $nF\phi$.

One may hope that any change in the electrode potential results in a similar change in the potential of the o.h.p., but this is not the case. The electrode potential is determined by the setting of a power supply, but $\phi_{o.h.p.}$ is determined by the solution composition, as well as by the setting of the power supply. Figure 21 illustrates this point in two extreme cases where (i) anions are adsorbed on a negatively charged surface and (ii) anions are adsorbed on a

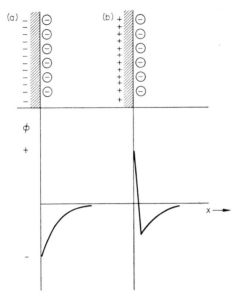

Fig. 21. Potential distance profiles: (a) Anions on a negatively charged electrode; (b) Anions on a positively charged electrode.

positively charged surface. In both cases the electroactive species is thought of as being a neutral molecule. It is seen that ϕ o.h.p. does not differ much from case (i) to case (ii) in spite of the extreme difference in electrode charge and potential.

(b) Since the interphase region is of very small thickness in not too dilute solutions, the field in it (field = gradient of potential) is very strong. A common value for the field strength is 10^8 volts m^{-1}. In this intense field many molecules and ions can be expected to be strongly polarized. Polarized species react in a different manner from unpolarized ones. This polarization may influence both the rate of the reaction (e.g., by bringing, say, the positive end of an induced dipole closer to a reducing electrode) and the mechanism (e.g., by orienting molecules in such a way as to bring a certain part of them close enough to the electrode).

(c) When ions are subjected to an electric field, their concentration at any point is modified according to the equation

$$c = c_0 \exp(-W/RT) \qquad 4.2$$

where c_0 is the concentration in the absence of field and W is the electrostatic work involved. Thus the concentration of ions in the double layer will differ from that in the bulk according to the field and the charge of the ions. Thus one expects that it will be difficult to reduce anions but easy to reduce cations on a negatively charged electrode.

A word of explanation may be needed here. It was said in Section B of Chapter 3 that migration of ions under the influence of electric field is not an important mass transfer mode in fairly concentrated solutions. Here we say that the field of the interphase changes the concentrations of ions. Are these statements not contradictory? That this is only an apparent contradiction becomes obvious when we remember that the thickness of the diffusion layer is commonly ten thousand times that of the double layer. Thus it is true that species move most of the distance from the bulk of the solution to the electrode surface in the gradient of forces other than electrical. At the same time the concentration at the electrode surface in the presence of field differs from that in the absence of one.

(d) As was stated before, many electrode reactions depend on the availability of suitable sites on the electrode (e.g. electrocrystallization and gas evolution). If the solution contains species that block these reaction sites, the reaction rate would drop. Sometimes, however, a different reaction path is taken, using the blocking molecules as catalysts, and the reaction rate increases. Thus the presence of species which are strongly adsorbed can reduce or enhance the rate of a given electrode process.

From the preceding discussion it is clear that to understand electrode reactions and to predict their behaviour at various conditions, we must know the potential distance curve in the interphase and its dependence on the solution constituents and potential of the electrode; we should know the influence of very strong electric fields on the reactant species and also the extent of adsorption of various species on the electrode. All this information could be derived readily had we known the microstructure of the interphase, i.e. the coordinates and velocities of all molecules present in the interphase region and the dependence of these coordinates and velocities on potential and other thermodynamic variables. This information, however, can never be obtained because of the interaction between the particles and our measuring device, an interaction which is formulated in Heisenberg's uncertainty principle. The next best data are the macrostructure, i.e. the concentration of all species as a function of potential, distance from the electrode and other variables. However, this information is also unavailable without the use of models because of the inadequacy of our experimental techniques.

There are two quantities which lend themselves to determination using only thermodynamic arguments and these are the charge on the electrode q_m and the surface excess Γ,

$$\Gamma_i = \int_0^\infty [c_i(x) - c_i^0]dx \qquad 4.3$$

c_i is the concentration at point x and c_i^0 is the concentration of i in the bulk. Here again we assume a unidimensional gradient of concentration. Figure 22

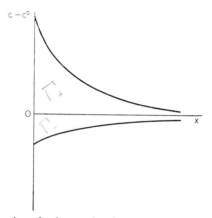

FIG. 22. Concentration of anions and cations at a negatively charged electrode.

shows a typical curve for the concentrations of anions and cations near a negatively charged electrode. The area between the upper curve and the x axis is Γ_+, the cation surface excess; the area between the lower curve and the x axis is Γ_-, the anion surface excess, which is negative.

The parameters which are of interest in the study of the interphase are the charge on the electrode, the concentration of the various solution species in the interphase and, consequently, the surface excess. As in all surface studies, these parameters are linked to the surface tension in a way which will be explained in Section C.

Models play a very important part in the study of the interphase. They give us ways to calculate the charge of the electrode as a function of electrode potential as well as complete potential-distance and concentration-distance curves. Models, however, are never ideal and those available for the interphase are quite crude. This must be borne in mind when using these models for quantitative calculations.

Let us now turn to look at the more popular models of the double layer first and then discuss the thermodynamic aspects of this field.

B. A quantitative treatment of the interphase

The quantitative treatment of the interphase will be presented in this section as it developed historically. This way we would become aware of the slow and steady development of the picture of the double layer as we understand it today and realize the complexity of the subject.

1. *The Helmholtz Model*

The first attempt to describe the electrode-solution interphase in electrostatic terms was made by Helmholtz in 1879. His model, which is shown in Fig. 23 is essentially that of a simple parallel plate capacitor: the charge on

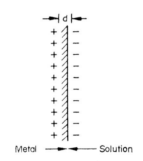

FIG. 23. The Helmholtz double layer.

the electrode is compensated by the ionic charge in solution, which is equal in magnitude and opposite in sign to the electrodic charge and is situated at a uniform distance, d, from the electrode surface. The capacity of this double layer is

$$C' = \frac{\varepsilon \Lambda}{d} \qquad 4.4$$

where C' is the capacity, ε is the permittivity of the solution in the double layer in units of metre^2 Newton^{-1} coulomb^{-2} and A is the area in metres^2. The potential drop V across this capacitor is

$$V = \frac{d}{\varepsilon A} q_m \qquad 4.5$$

and q_m is the charge on the electrode. In electrodics we are generally interested in quantities per unit area: current density, charge density and capacity per m². Therefore we may substitute in equations 4.4 and 4.5 $A = 1m^2$. The dielectric constant K is the ratio of the permittivity of the medium considered to that of free space, ε^0. Thus equations 4.4 and 4.5 become

$$C = \frac{K\varepsilon^0}{d} \qquad \text{4.4a}$$

$$V = \frac{d}{K\varepsilon^0}\, \sigma_m \qquad \text{4.5a}$$

where σ_m is the charge density and C the capacity per m². It is seen from this model that the capacity of the double layer is independent of potential, that the potential drop across the double layer should be linear with distance and that the charge should be a linear function of the inner potential of the electrode. Experimentally it is observed that the capacity is a function of potential and, as a result, the electrode charge is not a linear function of the potential. Thus the Helmholtz theory obviously needs refinement. Refinement and not rejection because there exists a separation of charges at the interphase, which resembles a capacitor in that these charges must be opposite in sign and equal in magnitude in order to preserve the electro-neutrality of the interphase.

2. The Gouy-Chapman Model

The next step in the development of the theory of the double layer is to remember that the ions are subject to thermal fluctuations as well as to the electric field. Gouy in France and Chapman in Britain solved this problem independently and the result is known as the Gouy-Chapman theory.

Consider Fig. 24, the electrode is of infinite area and the charges in solution are point charges. The term "infinite area" means that the area of the electrode is very much greater than the thickness of the double layer. Other

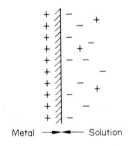

Metal — ►◄ — Solution

FIG. 24. The Gouy–Chapman double layer.

assumptions of this theory are that there are no dipoles at the electrode-solution interface (which means that the inner potential is equal to the outer potential) and that the potential ϕ and the charge density ρ are a function of x, the distance from the electrode, alone. Under these conditions the charge density is given by the Poisson equation

$$\rho = -K\varepsilon^0 \frac{d^2\phi}{dx^2} \qquad 4.6$$

which gives the influence of the electric field on the charge distribution. The effect of thermal fluctuations is given by the Boltzmann equation

$$n_i = n_i^0 \exp\left(-z_i e\phi/kT\right) \qquad 4.7$$

Here the number of particles (per unit volume) n_i of energy ε are given as a function of that energy. In an electric field, ε is the electrical energy $z_i e\phi$, z_i being the valence of the ion with sign, i.e. for Cl^- $z = -1$ and e the electronic charge in coulombs. n_i^0 is the number of particles per unit volume in the absence of electric field, i.e. far enough from the electrode so that $\phi = 0$. The charge density due to thermal fluctuations is

$$\rho = \sum_i n_i z_i e = \sum_i n_i^0 z_i e \exp\left(-z_i e\phi/kT\right) \qquad 4.8$$

Combining equations 4.6 and 4.8 yields

$$-K\varepsilon^0 \frac{d^2\phi}{dx^2} = \sum_i n_i^0 z_i e \exp\left(-z_i e\phi/kT\right) \qquad 4.9$$

If one substitutes $y = d\phi/dx$, then

$$\frac{d}{d\phi} y^2 = 2y \frac{dy}{d\phi} = 2 \frac{d\phi}{dx} \frac{dy}{d\phi} = 2 \frac{dy}{dx} = 2 \frac{d^2\phi}{dx^2} \qquad 4.10$$

therefore

$$\frac{1}{2} \frac{d}{d\phi} \left(\frac{d\phi}{dx} \right)^2 = -\frac{1}{K\varepsilon^0} \sum_i n_i^0 z_i e \exp\left(-z_i e\phi/kT\right) \qquad 4.11$$

Integration yields

$$\left(\frac{d\phi}{dx} \right)^2 = \frac{2kT}{K\varepsilon^0} \sum_i n_i^0 \exp\left(-z_i e\phi/kT\right) + \text{constant} \qquad 4.12$$

The constant can be evaluated by remembering that when $\phi = 0$ at the bulk of the solution $d\phi/dx = 0$ as well. Thus the final expression for the electric field is

$$\left(\frac{d\phi}{dx}\right)^2 = \frac{2kT}{K\varepsilon^0} \sum_i n_i^0 \left[\exp\left(-z_i e\phi/kT\right) - 1\right] \qquad 4.13$$

Let us now look at the special case of a single electrolyte in solution where $|z_+| = |z_-| = z$ and therefore $n_+^0 = n_-^0 = n^0$. For this case equation 4.13 becomes

$$\left(\frac{d\phi}{dx}\right)^2 = \frac{2kTn^0}{K\varepsilon^0} \left[\exp\left(-ze\phi/kT\right) - 1 + \exp\left(ze\phi/kT\right) - 1\right]$$

$$= \frac{2kTn^0}{K\varepsilon^0} \left[\exp\left(-ze\phi/kT\right) - 2\exp\left(-ze\phi/kT\right)\right.$$

$$\left. \times \exp\left(ze\phi/kT\right) + \exp\left(ze\phi/kT\right)\right]$$

$$= \frac{8kTn^0}{K\varepsilon^0} \sinh^2 ze\phi/2kT \qquad 4.14$$

Thus, taking square roots on both sides we get

$$\frac{d\phi}{dx} = -\left(\frac{8kTn^0}{K\varepsilon^0}\right)^{\frac{1}{2}} \sinh ze\phi/2kT \qquad 4.15$$

The negative sign of the square root was chosen because the potential at the bulk, i.e. when x is large, is zero. Thus, when ϕ is positive, its gradient is negative (ϕ decreases when x increases) and when ϕ is negative, the gradient is positive (ϕ increases, i.e. becomes less negative, as x increases).

Equation 4.15 gives the field at any point in the interphase as a function of the potential at that point. However, this is not the relationship we seek. We want to know the charge on the electrode and the capacity of the double layer as a function of its potential, the potential as a function of distance and the concentration of solution components as a function of the distance from the electrode. The first function will be found first, followed by the other functions.

The theorem used to find the charge on the electrode is known from electrostatics as Gauss's theorem. It states

$$\oint_s E.\vec{n}\, da = \frac{1}{K\varepsilon^0} \sum_{i=1}^{N} q_i \qquad 4.16$$

E is the vector electric field and the integral is taken over a closed surface S; da is an element of this surface, $\vec{n}$ is a vector of unit length and direction normal to the surface and the sum is taken over all charges. Putting equation

4.16 into words would be: the integral over a closed surface, of the field component normal to the surface, is equal to the charge inside this surface divided by the permittivity of the medium. The closed surface we use in the Gouy-Chapman theory is a drum, one face of which coincides with the electrode surface (Fig. 25). The sides of the drum are perpendicular to the

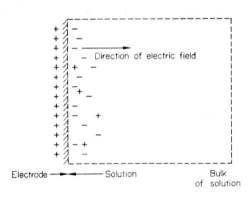

FIG. 25. A Gauss surface for the Gouy–Chapman theory; the broken line indicates the boundaries of the Gauss surface.

electrode surface and the other face is in the bulk of the solution. In this case the field in the bulk of the solution is zero and the component of the field normal to the sides of the drum is also zero. Thus we have only to consider the field at the electrode surface

$$\int_A \left(\frac{d\phi}{dx} \right)_{x=0} da = \frac{q}{K\varepsilon^0} \qquad 4.17$$

The field at the electrode surface is not a function of the exact location on the surface, i.e. it is constant for equation 4.17. The area of the electrode is unity, since we are interested in charge per unit area, so that equation 4.17 becomes

$$\left(\frac{d\phi}{dx} \right)_{x=0} = \frac{\sigma}{K\varepsilon^0} \qquad 4.18$$

Combining this result with equation 4.15 yields the desired function, but only for solutions containing one $z-z$ electrolyte:

$$\sigma = -(8kTK\varepsilon^0 n^0)^{\frac{1}{2}} \sin h\, ze\phi(x = 0)/2kT \qquad 4.19$$

for the charge of the solution side of the double layer. The charge on the electrode side is the same in magnitude and opposite in sign to the solution charge.

The capacity of the double layer σ/E is, obviously, not constant, but we can derive an expression for the differential capacity $C = d\sigma/dE$ and measure it as a function of electrode potential. As was said before, cells can be constructed so that the change in cell potential E is equal to the change in the electrode potential for any particular solution

$$C = \frac{d\sigma}{d\phi} = \left(\frac{2K\varepsilon^0 n^0 z^2 e^2}{kT} \right)^{\frac{1}{2}} \cosh \frac{ze\phi_0}{2kT} \qquad 4.20$$

If ϕ_0, the potential at the electrode surface, is small so that $ze\phi_0 \ll 2kT$, the hyperbolic cosine equals one and the capacity is

$$C = \left(\frac{2K\varepsilon^0 n^0 z^2 e^2}{kT} \right)^{\frac{1}{2}} \qquad 4.21$$

By comparing equations 4.21 and 4.4 we see that we can derive an equation for the thickness of the double layer, d

$$d = \left(\frac{kTK\varepsilon^0}{2n^0 z^2 e^2} \right)^{\frac{1}{2}} \qquad 4.22$$

This d is equal to the thickness of the ionic atmosphere in the Debye–Huckel theory for electrolytes for a $1:1$ electrolyte, it increases with decreasing concentration (n^0) and decreases with increasing concentrations. Thus the capacity of the double layer can also be written as

$$C = d \cosh \frac{ze\phi_0}{2kT} \qquad 4.23$$

Figure 26 shows curves calculated from equation 4.20 and some experimental results obtained with NaF solutions. (NaF is a suitable electrolyte

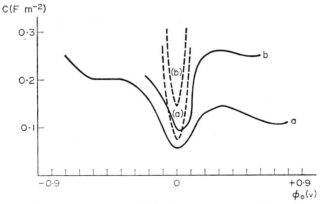

FIG. 26. Double layer capacity: – – – – Calculated from equation 4.20, ——— measured; the concentration increases from a to b.

for checking on simple theories because neither ion is specifically adsorbed on the mercury electrode). The discrepancy between the results is clear. Thus, the Gouy–Chapman theory, while being a considerable improvement over the Helmholtz theory in that it does not predict, for instance, constant capacities, is still rather unsatisfactory.

Before correcting the faults of the Gouy–Chapman model by using another model, let us derive expressions for the potential-distance and concentration-distance curves. In order to do that, equation 4.15 should be integrated.

Let us assume, for simplicity, that ϕ is very small, so that

$$\sinh \frac{ze\phi}{2kT} = \frac{ze\phi}{2kT}$$

If we allow a 10% deviation from the above expression, $ze\phi/2kT$ must be smaller than 0·73 (sin h 0·73 = 0·797). Thus the maximum value of ϕ for which this approximation is valid is somewhat less than 40 milivolts. In this case

$$\frac{\partial \phi}{\partial x} = - \left(\frac{8kTn^0}{K\varepsilon^0} \right)^{\frac{1}{2}} \frac{ze\phi}{2kT} = \left(\frac{2n^0 z^2 e^2}{K\varepsilon^0 kT} \right)^{\frac{1}{2}} \phi \qquad 4.24$$

and integration yields

$$\ln\phi = - \left(\frac{2n^0 z^2 e^2}{K\varepsilon^0 kT} \right)^{\frac{1}{2}} x + \text{constant} \qquad 4.25$$

The constant is simply ϕ_0, the potential at the surface of the electrode. Thus

$$\phi = \phi_0 \exp - \left(\frac{2n^0 z^2 e^2}{K\varepsilon^0 kT} \right)^{\frac{1}{2}} x \qquad 4.26$$

and the potential falls exponentially as the bulk of the solution is approached.

Equation 4.7 gives the concentration as a function of potential. Taking logarithms on both sides of equation 4.7 and combining with equation 4.26 yields

$$\ln n_{\pm} = \ln n^0 - \frac{z_{\pm} e\phi_0}{kT} \exp \left(\frac{2n^0 z^2 e^2}{K\varepsilon^0 kT} \right)^{\frac{1}{2}} x \qquad 4.27$$

The drop of concentration with distance is very rapid indeed.

3. The Stern Model

As was stated before, the Gouy–Chapman model is still too crude to provide satisfactory correlation between theory and experimental data. One obvious fault of this model is that it did not allow for the size of the ions, but considered them as point charges. Figure 27 shows a model of the double layer when the ions are no longer represented by their sign alone, but by small circles. The immediate result is that ions cannot approach the electrode to distances smaller than their radii and, therefore, the potential which they "feel" is not the electrode potential. The other result is that the potential-distance curve must now be considered in two parts, the potential drop up

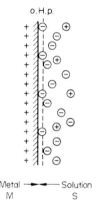

o.H.p.

Metal ——•◄—— Solution
   M   S

FIG. 27. The simple Stern model of the double layer.

to the location of the centres of the closest ions (the outer Helmholtz plane) and the potential drop from there to the bulk of solution.

$$^M\phi^s = {}^M\phi^{\text{o.h.p.}} + {}^{\text{o.h.p.}}\phi^s \qquad 4.28$$

By dividing all terms by σ_m, the charge density on the electrode, we get

$$\frac{^M\phi^s}{\sigma_m} = \frac{^M\phi^{\text{o.h.p.}}}{\sigma_m} + \frac{^{\text{o.h.p.}}\phi^s}{\sigma_m} \qquad 4.29$$

which can also be written as

$$\frac{1}{^MC^s} = \frac{1}{^MC^{\text{o.h.p.}}} + \frac{1}{^{\text{o.h.p.}}C^s} \qquad 4.30$$

This means that we can formally think of the double layer as composed of two capacitors connected in series. Changing notation equation 4.30 becomes

$$\frac{1}{C} = \frac{1}{C_i} + \frac{1}{C_d} \qquad 4.30\text{a}$$

C_i and C_d are the capacities of the inner "compact" part and the outer "diffuse" part of the double layer, respectively. C_d can be calculated from equation 4.20.

The Stern model can be regarded as the combination of the Helmholtz and the Gouy–Chapman models. C_i should thus correspond to the Helmholtz capacity and should be a constant for given species. Experimentally it is observed that C_i is fairly constant at a given potential over a range of concentrations of electrolytes where neither ion is specifically adsorbed. However, contrary to the model expectations, C_i varies considerably with potential. The Stern model does not explain this variation.

C_i at the potential of zero charge is computed in the following semiempirical, semi-theoretical way. The value of the capacity of the interphase is measured and the value of C_d is calculated from equation 4.20. Equation 4.30a is then used to calculate a value for C_i. Table IV gives values of C, C_d and C_i for various concentrations of NaF which were calculated in this way.

TABLE IV. *Capacities of the various parts of the double layer at the potential of zero charge*

Concentration M	Differential capacity ($\times 10^2$ F m^{-2})		
	C	C_d	C_i
0.001	6.0	7.2	32.00
0.01	13.1	22.8	30.79
0.1	20.7	72.2	29.02
1.0	25.7	228.0	28.96

Two things are evident from Table IV, the first is that the change in C_i over the given concentration range is much smaller than the change in C or in C_d; the second is that in concentrated solutions most of the capacity is accounted for by the capacity of the inner part of the double layer and, therefore, most of the potential drop occurs there. We use these deductions in order to test the validity of the Stern model, i.e. whether it is correct to divide the double layer into two parts which are related to each other like capacitors in series.

In order to calculate C_i as a function of potential we need information on C_d, σ_m and ϕ_0 (which now is the potential at the plane of closest approach and not at the electrode surface). σ_m is evaluated by integrating the C vs. potential curve and ϕ_0 is found from equation 4.19. Equation 4.20 is used next to calculate C_d. (a) C_i is now calculated from equation 4.30a. The resulting curve is given in Fig. 28. Once this curve is known, we can reverse this process and calculate C for a different concentration of the same electrolyte. (b) If the experimental curve fits the calculated one, it is deduced that the curve C_i vs. potential does not change as the concentration of the electrolyte changes, i.e. the division of the capacity to "compact" and "diffuse" parts is correct. The capacity-potential curve calculated in this way for $0 \cdot 01$ M NaF solutions, together with the experimental values are given in Fig. 29. The agreement between the two curves is quite good, except in the range where the electrode has a high positive charge. It is known, however, that in this region fluoride ions are specifically adsorbed on mercury, invalidating the assumption basic to these calculations.

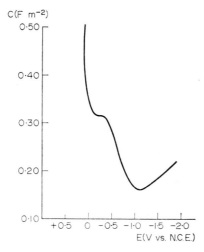

FIG. 28. Capacity of the inner double layer in NaF.

It has been established in this section that it is correct to regard the interphase as two capacitors in series and that the capacity of the inner layer is characteristic of the electrolyte present, and is independent of concentration, as expected from the model. However, the Stern model does not explain the

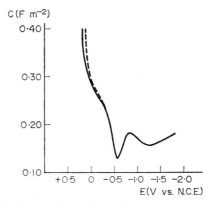

FIG. 29. Calculated and experimental capacities for $0 \cdot 01$ M NaF.

shape of the C_i vs. potentential curve, nor does it say anything about the factors which influence specific adsorption.

4. *The Modern Model of the Interphase*

There are three points related to the structure of the interphase that the models described so far do not consider at all: (*a*) the value of the distance of closest approach, (*b*) the influence of hydration (or solvation) on the interphase and (*c*) the value of the permittivity (or dielectric constant) in the inner double layer.

A careful analysis of the first mentioned factor will result in the understanding of the other two as well. We shall therefore try now to find the best value of the distance of closest approach which agrees with experimental facts, and appropriately begin by describing these facts. Figure 30 shows

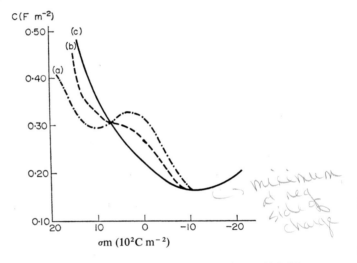

Fig. 30. Capacity of the inner double layer; (*a*) 0°C, (*b*) 45°C, (*c*) 85°C.

the capacity of the inner part of the double layer plotted against the charge on the electrode at several temperatures in NaF solutions. Two features of these curves should be noted—the minimum at the negative side of charge and the "hump" at the positive side. The remarkable feature of the minimum in the capacity-charge curve is that it is constant and independent of temperature, concentration, nature of the anions and the nature of the cations in solution, as shown in Table V.

The "hump" on the other hand, is dependent on many variables: its height is dependent on temperature and its location, as well as height, depends on the concentration and nature of the solution constituents.

The independence of this minimum in the C_i vs. σ_m curve of the nature of anions can be readily explained by remembering that at the high negative charge ($\sigma_m = -12 \times 10^{-2} \, C \, m^{-2}$) anions are repelled from the electrode and the layer closest to the metal surface is that of cations. The independence of temperature shows that the structure of the inner part of the interphase is

TABLE V. *Values for C_i for solutions of various cations at $\sigma_m = -0.12 \, C \, m^{-2}$*

Ion	Unhydrated radius $\times 10^{10} \, m$	Hydrated radius $\times 10^{10} \, m$	Differential capacity $\times 10^2 \, F \, m^{-2}$
H^+	—	—	16·6
Li^+	0·60	2·9	16·2
K^+	1·33	4·1	17·0
Rb^+	1·48	3·6	17·5
Mg^{2+}	0·65	4·1	16·5
Sr^{2+}	1·13	4·0	17·0
Al^{3+}	0·50	4·8	16·5

so very distorted by the strong electric field that the thermal energy is negligible. This implies that the dielectric constant of the solvation layer immediately adjacent to the surface of the metal differs considerably from that of bulk solvent and rather resembles the dielectric constant of the first solvation layer of ions. It is the independence of this minimum from the nature of cations that leads us to the value of the distance of closest approach. Since ions are usually hydrated in solution and the electrode also has a layer of water attached to it by similar forces, there are three possible values for the distance of closest approach: (a) unsolvated ions in contact with unsolvated electrode, (b) solvated ions in contact with unsolvated electrode (or the other way round) and (c) solvated ions in contact with solvated electrode.

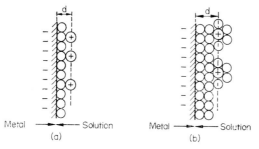

FIG. 31. The two possibilities for the distance of closest approach, d; (a) Bare cations to hydrated electrode, (b) Hydrated cations to hydrated electrode.

The first possibility is ruled out immediately; from the independence of the capacity minimum of the nature of the cations one must conclude that the distance of closest approach for all cations should be approximately the same. But the radii of the bare cations differ from each other considerably! Let us now consider the two other possibilities by referring to Fig. 31, which shows the distances of closest approach for these two conditions. If r_i denotes the radius of the bare ion and r_w the radius of the water molecule, the distance of closest approach, d, for possibility (b) is $2r_w + r_i$ and the capacity is

$$C_i = \frac{K\varepsilon^0}{d} = \frac{K\varepsilon^0}{r_i + 2r_w} \qquad 4.31$$

Table VI shows values calculated for the ions in Table V for a dielectric constant of 6, which was found to be the dielectric constant of the primary hydration layer. ε^0, the permittivity of free space is $8 \cdot 854 \times 10^{-12}\, C^2\, N^{-1}\, m^{-2}$ and the radius of a water molecule is $1 \cdot 54 \times 10^{-10}\, m$.

TABLE VI. *Calculated values for C_i from equation* 4.31

Ion	r_i $\times 10^{10}$ m	C_i $\times 10^2\, F\, m^{-2}$
Li^+	0·60	14·43
K^+	1·33	12·05
Rb^+	1·48	11·65
Mg^{2+}	0·65	14·24
Sr^{2+}	1·13	12·62
Al^{3+}	0·50	14·83

The values calculated are too low and depend more on the radius of the cation than the experimental values.

The possibility (c) involves two layers of water molecules between the first layer of cations and the metal. The first layer of water can be regarded as a hydration layer with a dielectric constant of 6. The second layer is not uniform, some of the water there will be water of hydration of cations (dielectric constant 6) but some of it will not be particularly associated with any ion. The dielectric constant of unassociated water in this second layer should be between 6 and the bulk value of 80; one estimate may be 35. In not too concentrated solutions most of the second layer will be composed of unassociated water molecules, and, thus, one may assume that the "average" dielectric constant of this second layer of water is 35. If we have a capacitor which is filled with two sheets of different dielectric material we may regard it as two capacitors in series, one with one dielectric and the other with the second. Thus

$$\frac{1}{C} = \frac{d_1}{\varepsilon_1} + \frac{d_2}{\varepsilon_2} \qquad\qquad 4.32$$

From Fig. 31b

$$\frac{1}{C_i} = \frac{2r_w}{6\varepsilon^0} + \frac{2r_w + r_i}{35\varepsilon^0} \qquad\qquad 4.33$$

Table VII gives the results of calculations made from equation 4.33.

TABLE VII. *Calculated values for C_i from equation 4.33*

Ion	r_i $\times 10^{10}$ m	C_i $\times 10^2$ F m^{-2}
Li$^+$	0·60	14·31
K$^+$	1·33	13·85
Rb$^+$	1·48	13·76
Mg^{2+}	0·65	14·28
Sr^{2+}	1·13	13·97
Al^{3+}	0·50	14·38

Comparing the calculated value of C_i in Table VII to the experimental ones in Table V shows that the calculated values are constant and independent of cationic dimensions. There are, however, two points for which the calculated and experimental values do not agree: (a) the calculated values are too low and (b) small ions have small experimental capacity values, but larger calculated capacity values. The reason for this disagreement is neglect of several factors in the preceding discussion as follows. The thickness of the second capacitor d_2 was taken to be the radius of the ion plus the diameter of a water molecule. This approach is over-simplified, hydrated ions may have a variety of radii (Table V) which are not necessarily $r_i + 2r_w$. The first water layer at the electrode surface is strongly polarized, giving rise to additional capacity. The electrical field itself exerts a force on the molecules and ions, changing somewhat the packing at the interphase. The charges at the interphase are not smeared out but are discrete with uncharged spaces between them.

Since the object of this section is to give a semiquantitative view of the interphase structure, and we are able to account for the constancy of C_i, calculating values which are not very far from the experimental ones for the minimum in the capacity potential curve, we shall leave this point here, and turn to look more closely at ways and methods for treating adsorption on electrodes.

C. Adsorption from solution

This section will survey adsorption phenomena of ions and neutral molecules on the surface of mercury in contact with aqueous solutions at equilibrium. We shall assume throughout that no electrochemical reaction takes place at the interphase under discussion and, therefore, use thermodynamic methods for the calculation of the desired parameters. The parameters that are of interest in the study of adsorption are the surface excess Γ and its dependence on the electrode potential and concentration of adsorbate in the bulk; the dependence of the charge of the electrode on its potential is also of interest. We shall, therefore proceed to discuss that part of thermodynamics which is relevant to adsorption on electrodes and then consider adsorption of ions first and of neutral molecules second.

1. The Gibbs Adsorption Isotherm

The change of internal energy ΔU of a system is equal to q, the heat supplied minus w the work done by the system.

$$\Delta U = q - w \qquad 4.34$$

For a thermodynamically reversible process

$$\Delta S = q/T \qquad 4.35$$

where ΔS is the entropy change and T the temperature. The work done by the system includes pressure-volume (P, V), increase in surface area (A) and potential dependent terms as well as a term w_i, the work done by changing the composition of the system.

$$w = P\Delta V + \gamma\Delta A + \phi\Delta q_m + w_i \qquad 4.36$$

γ is the surface tension, ϕ the inner potential and q_m the total charge of the electrode. Considering an infinitesimal process we write

$$dU = TdS - (PdV + \gamma dA + \phi dq_m + w_i) \qquad 4.37$$

We define now the chemical potential

$$\mu_i = \left(\frac{\partial U}{\partial n_i} \right)_{S, V, A, q_m, n_{j \neq i}} \qquad 4.38$$

So that $w_i = -\sum \mu_i \, dn_i$. Equation 4.37 now becomes

$$dU = TdS - PdV - \gamma dA - \phi dq_m - \sum \mu_i \, dn_i \qquad 4.39$$

The Gibbs free energy function G is defined as

$$G = U + PV - TS \qquad 4.40$$

$$dG = dU + PdV + VdP - TdS - SdT \qquad 4.41$$

Substituting dU from equation 4.39 one gets

$$dG = VdP - SdT - \gamma dA - \phi dq_m - \sum \mu_i \, dn_i \qquad 4.42$$

Integration of equation 4.39 yields†

$$U = TS - PV - \gamma A - \phi q_m - \sum \mu_i n_i \qquad 4.43$$

Therefore by equation 4.40

$$G = -\gamma A - \phi q_m - \sum \mu_i n_i \qquad 4.44$$

$$dG = -\gamma dA - Ad\gamma - \phi dq_m - q_m d\phi - \sum(\mu_i \, dn_i + n_i \, d\mu_i) \qquad 4.45$$

Since equations 4.42 and 4.45 must be equal, one gets

$$VdP - SdT - Ad\gamma - q_m d\phi - \sum n_i \, d\mu_i = 0 \qquad 4.46$$

which for $T, P =$ constant yields

$$Ad\gamma = -q_m d\phi - \sum n_i \, d\mu_i \qquad 4.47$$

Dividing equation 4.47 by A gives

$$d\gamma = -\sigma_m d\phi - \sum(n_i/A)d\mu_i \qquad 4.48$$

which is one form of the Gibbs adsorption isotherm. It remains to see now what is the relationship between n_i/A and the surface excess Γ_i as defined above

$$\Gamma_i = \int_0^\infty (c_i - c_i^0)dx$$

The concentration, ch, of any species is the amount (or the number of molecules) divided by the volume occupied by this amount $hc = n/V$. dx can be related to the volume by

$$dx = \frac{1}{A} dv$$

† The method used for the integration follows from the basic definition of the integral as a limit of a sum and corresponds to expanding the system, keeping T, P and relative proportions of the components constant. The exact way in which this is done is given in textbooks on thermodynamics.

Hence

$$\Gamma_i = \frac{1}{A} \int_0^\infty (c_i - c_i{}^0) dV = \frac{n_i}{A} - \frac{n_i{}^0}{A} \qquad 4.49$$

$$\sum (n_i/A) d\mu_i = \sum \left(\Gamma_i + \frac{n_i{}^0}{A} \right) d\mu_i \qquad 4.50$$

The Gibbs–Duhem relation at constant temperature and pressure†

$$\sum n_i{}^0 d\mu_i = 0 \qquad 4.51$$

and, therefore, equation 4.48 becomes

$$d\gamma = -\sigma_m d\phi - \sum \Gamma_i d\mu_i \qquad 4.52$$

This is the form of the Gibbs adsorption isotherm that is commonly used. It relates changes in the surface tension γ to the charge density and the surface excess

$$\left(\frac{\partial \gamma}{\partial \phi} \right)_{\mu_i} = -\sigma_m \qquad 4.53$$

$$\left(\frac{\partial \gamma}{\partial \mu_i} \right)_{\phi, \mu_{j \neq i}} = -\Gamma_i \qquad 4.54$$

The potential appearing in equation 4.53, ϕ, is the potential between the electrode and the bulk of solution, a quantity which is not experimentally accessible. In order to derive experimentally meaningful equations, let us introduce a reference electrode whose potential ϕ_{re} is determined by species in the cell solution. The cell potential E is

$$E = \phi - \phi_{re} \qquad 4.55$$

$$dE = d\phi - d\phi_{re} \qquad 4.56$$

Equation 4.52 thus becomes

$$d\gamma = -\sigma_m dE - \sigma_m d\phi_{re} - \sum \Gamma_i d\mu_i \qquad 4.57$$

Since ϕ_{re} does not depend on the charge density σ_m of the working electrode equation 4.53 is valid also in the form

$$\left(\frac{\partial \gamma}{\partial E} \right)_{\mu_i} = -\sigma_m \qquad 4.58$$

† The Gibbs–Duhem relation in this form is valid only when electric fields and surface forces are absent. The equivalent expression for systems involving electric fields and surfaces is

$$-Ad\gamma - qd\phi + \Sigma n_i d\mu_i = 0$$

Therefore

$$\Sigma n_i d\mu_i \neq 0$$

The discussion of the equivalent equation for the determination of the surface excess is postponed until later.

The differential capacity of the double layer is simply given by the derivative of σ_m

$$C = \left(\frac{\partial \sigma_m}{\partial E}\right)_{\mu_i} = -\left(\frac{\partial^2 \gamma}{\partial E^2}\right)_{\mu_i} \qquad 4.59$$

We have seen that the electrochemical variables, σ_m, C and Γ are all related to the surface tension of the electrode. Measurements of surface tension as a function of electrode potential and solution concentration, thus play a very important part in the study of the double layer. It is obviously quite simple to measure surface tension of a liquid electrode and very difficult to do the same with solid ones. Mercury once more has a decisive advantage over all other metals for the study of the interphase at around room temperature.

It is convenient to divide the discussion of adsorption, at this point, to adsorption of ions and that of neutral molecules. The reason for this is that ions are present in the solution in large concentrations, adsorbable ions are often constituents of the supporting electrolyte and they also determine the potential of the reference electrode. On the other hand the concentration of neutral species is usually kept low to ensure that they have no effect on the reference electrode.

2. Adsorption of Ions

The first point to be taken up in this section is the derivation, in terms of cell potential, of an alternative equation to 4.54. In order to do this let us consider a solution with only one electrolyte of univalent ions. Denoting the surface excess, chemical potential etc. of the cation by subscript $+$ and the same of the anion by subscript $-$, the chemical potential of this electrolyte is

$$\mu = \mu_+ + \mu_- \qquad 4.60$$

$$d\mu = d\mu_+ + d\mu_- \qquad 4.61$$

assuming the anion determines the potential of the reference electrode:

$$d\phi_{re} = -d\mu_-/n\text{F} \qquad 4.62$$

Then for a constant cell potential

$$-d\gamma = \frac{\sigma_m}{n\text{F}} d\mu_- + \Gamma_+ d\mu_+ + \Gamma_- d\mu_- \qquad 4.63$$

Substitution of $d\mu_+$ from equation 4.61 gives

$$dy = -\frac{\sigma_m}{nF} d\mu_- - \Gamma_+ d\mu + \Gamma_+ d\mu_- - \Gamma_- d\mu_-$$

$$= -\Gamma_+ d\mu + \left(\Gamma_+ - \Gamma_- - \frac{\sigma_m}{nF} \right) d\mu_- \qquad 4.64$$

Since the interphase as a whole is neutral, the sum of the charges comprising it must be zero. In equation form

$$\sigma_m + nF\Gamma_+ - nF\Gamma_- = 0 \qquad 4.65$$

Equation 4.64 thus becomes

$$dy = -\Gamma_+ d\mu \qquad 4.66$$

at a constant cell potential, or

$$\left(\frac{\partial y}{\partial \mu} \right)_E = -\Gamma_+ \qquad 4.67$$

Thus the change of surface tension with concentration at constant cell potential is equal to the surface excess of the cation when the reference electrode used is reversible toward the anion and vice versa.

In the following we shall discuss the way by which ionic surface excesses are usually determined and specific adsorption is established and measured. When using mercury electrodes, data on surface tension can be obtained directly using techniques to be described in Chapter 5. These data are in the form of surface-tension vs. potential curves, and a series of these is collected in order to determine Γ. The standard procedure is as follows: making sure that the reference electrode used is reversible to the cations in solution if Γ_- is sought, the surface tension is plotted vs. $\ln a$ (a being the activity of the electrolyte) at constant cell potential.

$$-\Gamma_- = \frac{1}{RT} \left(\frac{\partial y}{\partial \ln a} \right)_E \qquad 4.68$$

The graphical differentiation of these curves gives Γ_- directly. This is not the surface concentration of specifically adsorbed anions but the surface excess. In other words: Γ_- gives us the effect of the electric field and the specific forces combined. Usually we are interested in the surface excess due to specific adsorption alone and therefore we must subtract from Γ_- that part of the surface excess due to coulombic forces. The change density associated with Γ_- is

$$\sigma_- = z_- F\Gamma_- \qquad 4.69$$

where z_- is the "valence" of the anions. The charge density on the metal is given by equation 4.58

$$\left(\frac{\partial \gamma}{\partial E} \right)_{\mu_1} = -\sigma_m$$

Therefore, the cationic charge density is

$$\sigma_+ = \sigma_m - \sigma_- \qquad 4.70$$

We now calculate the potential at the outer Helmholtz plane from the modified Gouy–Chapman theory remembering that ϕ_0 is no longer the potential at the electrode surface but, according to the Stern model, that of the o.h.p. q in equations 4.17 et seq. no longer equals the negative of the electrode charge but is that of the ions in the diffuse part of the interphase, i.e. those ions which are not specifically adsorbed. In order to determine ϕ_0 we must consider for the time being only that part of the surface excess which is due to coulombic forces $\Gamma'_{\pm}$, we write from the definition of Γ and equation 4.7

$$\Gamma'_+ = \frac{n^0}{N} \int_{\text{o.h.p.}}^{\infty} \left[\exp\left(-\frac{z_+ e\phi_0}{kT} \right) - 1 \right] dx \qquad 4.71$$

$$\Gamma'_- = \frac{n^0}{N} \int_{\text{o.h.p.}}^{\infty} \left[\exp\left(-\frac{z_- e\phi_0}{kT} \right) - 1 \right] dx \qquad 4.72$$

where N is Avogadro's number. From equation 4.69

$$\sigma'_+ = \frac{z_+ Fn^0}{N} \int_{\text{o.h.p.}}^{\infty} \left[\exp\left(-\frac{z_+ e\phi_0}{kT} \right) - 1 \right] dx \qquad 4.73$$

$$\sigma'_- = \frac{z_- Fn^0}{N} \int_{\text{o.h.p.}}^{\infty} \left[\exp\left(-\frac{z_- e\phi_0}{kT} \right) - 1 \right] dx \qquad 4.74$$

After fairly extensive manipulation (see Appendix at the end of this Chapter) of these equations and the use of equation 4.19 we get

$$\sigma'_+ = (2kT K\varepsilon^0 n^0)^{\frac{1}{2}} \left[\exp\left(-\frac{z_+ e\phi_0}{2kT} \right) - 1 \right] \qquad 4.75$$

and similarly

$$\sigma'_- = (2kT K\varepsilon^0 n^0)^{\frac{1}{2}} \left[\exp\left(-\frac{z_- e\phi_0}{2kT} \right) - 1 \right] \qquad 4.76$$

where z is taken with sign, i.e. z for Cl^- is -1.

If we assume no specific adsorption of cations $\sigma_+ = \sigma'_+$ and equation 4.75 can be used to calculate ϕ_0. This in turn can be used in equation 4.76 to calculate σ'_-. The charge due to specific adsorption is $\sigma_- - \sigma'_-$, or the surface excess due to specific adsorption Γ_{sa} is

$$\Gamma_{sa} = \frac{\sigma_- - \sigma'_-}{nF} \qquad\qquad 4.77$$

Table VIII compares experimental values of σ_m and σ_{sa} ($\sigma_{sa} = nF\Gamma_{sa}$) from $1 \cdot 0\,M$KCl solutions.

TABLE VIII. *Comparison of electrode charge with the specifically adsorbed charge*

σ_m C m^{-2}	σ_{sa} C m^{-2}
$+0\cdot20$	$-0\cdot301$
$+0\cdot16$	$-0\cdot247$
$+0\cdot12$	$-0\cdot201$
$+0\cdot08$	$-0\cdot157$
$+0\cdot04$	$-0\cdot111$
$+0\cdot00$	$-0\cdot065$
$-0\cdot04$	$-0\cdot026$
$-0\cdot08$	$-0\cdot005$
$-0\cdot12$	$0\cdot000$
$-0\cdot16$	$+0\cdot001$
$-0\cdot20$	$+0\cdot006$

Since $d\gamma$ is a complete differential, it follows from the Gibbs adsorption isotherm that

$$\left(\frac{\partial E}{\partial \mu_i} \right)_q = - \left(\frac{\partial \Gamma i}{\partial q} \right)_\mu \qquad\qquad 4.78$$

This equation implies that the potential of zero charge changes when the concentration of adsorbed species changes. Figure 32 shows electrocapillary curves in various concentrations of aqueous HCl showing very well the shift in the electrocapillary maximum (which is the potential of zero charge). This phenomenon serves as a criterion for the existence of specific adsorption, whenever the potential of zero charge changes with concentration of species i, that species is specifically adsorbed. However, the reverse does not hold, when the p.z.c. is constant with concentration of species i, that species can still be specifically adsorbed at high electrode charges. The best example to illustrate this point is the fluoride ion which had been used for many years as a standard for interphase studies because the potential of zero charge in its solutions does not change with concentration (showing, presumably, the lack of its specific adsorption). It was later discovered that at high positive

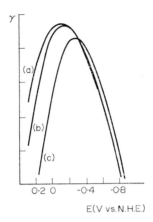

FIG. 32. Electrocapillary curves for HCl in various concentrations; (a) $0 \cdot 01\ M$ (b) $0 \cdot 10\ M$, (c) $1 \cdot 0\ M$.

electrode charge and high concentrations, fluoride ion is specifically adsorbed on the mercury electrode. It is still true that at the point of zero charge fluoride is not specifically adsorbed.

Why does specific adsorption take place, what are the forces involved in it and what are the factors which influence it? We shall start the discussion by considering the energetics of specific adsorption, starting with species A in solution and ending with species A adsorbed on the electrode surface. In order to go through this process we have first to strip A of its hydration layer, investing the hydration energy ΔG_H. Then we must strip a portion of the electrode of its hydration layer which involves the energy ΔG_{HE}. A can then be allowed to adsorb on the electrode with the gain of the energy of adsorption ΔG_A. A will be partially hydrated when adsorbed; the energy of partial hydration being ΔG_{PH}. If specific adsorption takes place, the sum of all these energies is negative

$$\Delta G_H + \Delta G_{HE} - \Delta G_A - \Delta G_{PH} < 0 \qquad 4.79$$

ΔG_{HE} is a property of the electrode and not of the adsorbing species and $\Delta G_H - \Delta G_{PH}$ is a fraction of the energy of hydration and is always positive. Thus energetically the phenomenon of specific adsorption is reduced to the difference between a certain fraction of the energy of hydration and the energy of adsorption; if the first is greater then there is no specific adsorption but if the latter is greater, specific adsorption is observed. Table IX gives values for the free energy of hydration of the halide anions together with the bond strengths of the Hg–X bond in the compounds Hg_2X_2 and their adsorbability (given as σ_{sa} when $\sigma_m = 0 \cdot 10$ C m^{-2}).

TABLE IX. *Energy of hydration, bond strength and adsorbability of the halogen anions*

	$-\Delta G_H$ k J mole^{-1}	Bond strength k J mole^{-1}	Adsorbability C m^{-2}
F$^-$	476	134	0·095
Cl$^-$	371	96·3	0·146
Br$^-$	326	71·2	0·192
I$^-$	293	28·3	0·292

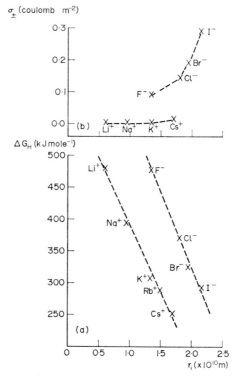

FIG. 33. Energy of hydration and adsorbability of some ions; σ_- is given at $\sigma_m = 0 \cdot 10$ C m^{-2}, σ_+ is given at $\sigma_m = -0 \cdot 15$ C m^{-2}.

It is quite clear that the adsorbability increases as the hydration energy decreases; however, surprisingly, the adsorbability does not increase when the bond strength increases, quite the contrary. This shows that no actual chemical bonds are formed in specific adsorption. The close association between adsorbability and energy of hydration which, in turn is connected with the radius of the ion, is shown in Fig. 33. It is evident that anions in general adsorb much more than cations and that adsorbability of both types of ions increases with their radii.†

Figure 34 shows the dependence of the specifically adsorbed charge of chloride and iodide on the activity of the electrolyte solution, i.e. it shows the adsorption isotherms. It is seen that even these anions exhibit two very different curves. Several theories were proposed in order to explain these and other experimental data, but none has become generally acceptable. Figure 35 shows the dependence of the specifically adsorbed charge from various $0 \cdot 1\, M$ electrolyte solutions on the charge of the electrode; again a satisfactory theoretical explanation for these curves is lacking.

It is of great interest to calculate the potential at the o.h.p. for various electrolytes; the o.h.p. is the site where many electrode reactions are thought

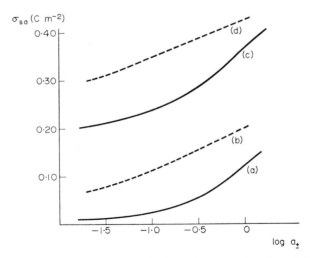

FIG. 34. Amount of specifically adsorbed Cl^- and I^- as a function of their activity; (a) $\sigma_m = 0$, KCl; (b) $\sigma_m = 0$, KI; (c) $\sigma_m = 0 \cdot 18\, C\, m^{-2}$, KCl; (d) $\sigma_m = 0 \cdot 18\, C\, m^{-2}$, KI.

† The values of standard free energy of hydration for electrolytes depends on the standard states chosen for the solution and for the gas phase; therefore they will differ between workers who use different standard states. Moreover, the values of standard free energy of hydration of ions depends entirely on the convention used for splitting up the values for the electrolytes. Thus the absolute values used in Fig. 33(a) are not significant; it is their relative trend which is significant.

to take place. The method of calculation was shown above and some results for $0\cdot1\,M$ solutions are given in Fig. 36.   It is clearly seen that except for the fluoride salt, ϕ_0 is never positive; all anions adsorb specifically in such a way as to more than compensate for the positive charge on the electrode.

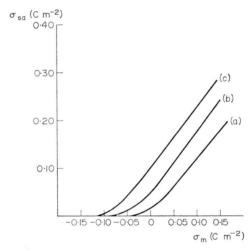

FIG. 35. Amount of specifically adsorbed anions as a function of charge density of the metal; (a), $0\cdot1\,M$ KCl, (b) $0\cdot1\,M$ KBr, (c) $0\cdot1\,M$ KI.

Thus as a rule one may think of the potential at the o.h.p. in almost all electrolyte solutions of this and higher concentrations as negative when considering the effect of the interphase structure on electrode processes.

We have seen in this section that information on the surface excess of ions is experimentally available, provided one uses the correct reference electrode.   We illustrated the correlation between energy of hydration of ions and their adsorbability, we also established that specific adsorption is probably not due to formation of chemical bonds.   We saw that specific adsorption of anions is quite general, but cations adsorb only if they are rather large; the extent of adsorption and its dependence on the solution activity and the electrode potential have remained without a generally accepted explanation.

3. Adsorption of Neutral Molecules

Most studies of adsorption of neutral molecules are concerned with organic molecules (any substance which adsorbs at a surface is called "surface active" substance).   These are usually present in only a small concentration in the electrochemical cell in solutions which contain fairly high concentrations of electrolytes. Therefore the potentials of the reference electrode and the liquid junction are not influenced by the presence of these organic particles and any change in the electrode potential ϕ, is reflected in the cell potential E $(d\phi = dE)$.

$$\sigma_m = - \left(\frac{\partial \gamma}{\partial \phi} \right)_{\mu_i} = - \left(\frac{\partial \gamma}{\partial E} \right)_{\mu_i} ;$$

$$\Gamma_i = - \left(\frac{\partial \gamma}{\partial \mu_i} \right)_{\phi, \mu_{j \neq i}} = - \left(\frac{\partial \gamma}{\partial \mu_i} \right)_{E, \mu_{j \neq i}} \qquad 4.80$$

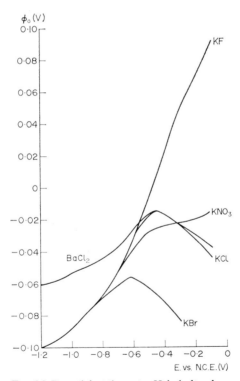

FIG. 36. Potential at the outer Helmholtz plane.

In the case considered here, all adsorption is regarded as specific adsorption. Our simple models of the interphase do not allow the calculation of surface excess due to dipole-electrode-field interaction; pure coulombic, i.e. charge-electrode-field interaction is, of course, absent in neutral molecules.

Let us now consider adsorption of molecules whose dipole moment is not very large (smaller than that of water) and which are not polarizable. If the electrode is either highly positive or highly negative, it is expected that the polar components of the solution, i.e. water and ions would adsorb very strongly, leaving no adsorption sites for the neutral molecules. At the point of zero charge the interaction between the polar components and the electrode

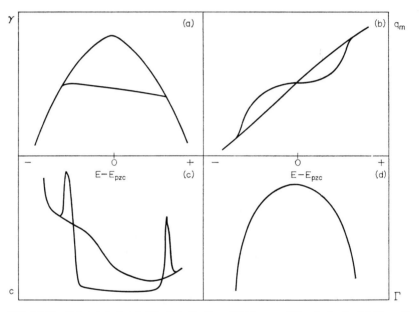

FIG. 37. Typical curves for: (a) Surface tension; (b) Charge of the electrode; (c) Capacity and (d) surface excess of adsorbed neutral species.

will be minimum and, as a result, adsorption of neutral molecules will be maximum. Thus adsorption of neutral molecules takes place only at potentials not too far from the potential of zero charge. Figure 37a shows a schematic presentation of a typical shape of the surface-tension-potential curve with and without the presence of specifically adsorbed species. From the Gibbs adsorption isotherm, curves for the electrode charge and surface excess can be readily obtained. Schematic curves for these and for the capacity of the interphase are also shown in Fig. 37. The capacity potential curve is of particular interest as it exhibits two very sharp maxima at the adsorption (or desorption) potentials on the two sides of the point of zero charge. The height of these maxima depends on the frequency used to measure the capacity, as can be shown from the following argument. Since

dq_m is a complete differential

$$dq_m = \left(\frac{\partial q_m}{\partial E} \right)_\Gamma dE + \left(\frac{\partial q_m}{\partial \Gamma} \right)_E d\Gamma \qquad 4.81$$

it follows that

$$C = \left(\frac{\partial q_m}{\partial E} \right)_\mu = \left(\frac{\partial q_m}{\partial E} \right)_\Gamma + \left(\frac{\partial q_m}{\partial \Gamma} \right)_E \left(\frac{\partial \Gamma}{\partial E} \right)_\mu \qquad 4.82$$

At low frequencies the measured C is given by equation 4.82, but at high frequencies the rate of adsorption and desorption becomes too slow to follow exactly the changes in the electrode potential. The value of $(\partial \Gamma / \partial E)_\mu$ falls as the frequency is increased becoming eventually zero, so that at high frequency the capacity is given by

$$C_\infty = \left(\frac{\partial q_m}{\partial E} \right)_\Gamma \qquad 4.83$$

In general the point of zero charge in the presence of adsorbed species is different from that in its absence. This change may be positive, as observed for several alphatic compounds, or negative, as observed with some aromatic compounds. Opinions on the cause of this phenomenon vary, but it seems that this change of the p.z.c. is due to changes in χ, the surface potential, which arise because of the dipole moment of the adsorbate and its interaction with the conductance band of the electrode.

The shapes of the charge-potential and the capacity-potential curves follow from the Gibbs adsorption isotherm immediately. The shape of the surface-excess-potential curve is more difficult to derive. This will be the subject to be discussed next and in order to do that we shall consider the general adsorption isotherm

$$Bc = f(\theta) \qquad 4.84$$

where c is the concentration of adsorbed species in the bulk and θ is the fraction of the electrode surface covered with adsorbate. B is a function of the electrode potential, $B = B(\phi)$. Alternatively the adsorption isotherm can be written as

$$Hc = h(\theta) \qquad 4.85$$

where H is a function of the electrode charge, $H = H(q)$. B and H are the adsorption equilibrium constants referred to in section $A\,2$ of Chapter 3. Whether equation 4.84 or 4.85 is used depends on the type of adsorption data and on the point of view of the user. We shall proceed using equation 4.84. Taking logarithms of both sides and differentiating with respect to

ϕ yields

$$\frac{d\ln B}{d\phi} = -\left(\frac{\partial \ln c}{\partial \phi}\right)_\theta \qquad 4.86$$

With this equation in mind, let us turn now to the Gibbs adsorption isotherm written in a different form

$$d\gamma = -\sigma_m\, d\phi - A\,\theta\, d\ln c \qquad 4.87$$

where $A = RT\Gamma_m$, Γ_m being the maximum surface excess. $d\ln c$ appears because

$$d\mu = RT\, d\ln c \qquad 4.88$$

It follows from equation 4.87 and the property of $d\gamma$ being a complete differential

$$\left(\frac{\partial \sigma_m}{\partial \ln c}\right)_\phi = A\left(\frac{\partial \theta}{\partial \phi}\right)_c \qquad 4.89$$

$$\left(\frac{\partial \ln c}{\partial \phi}\right)_\theta = -\left(\frac{\partial \ln c}{\partial \theta}\right)_\phi \left(\frac{\partial \theta}{\partial \phi}\right)_c \qquad 4.90$$

Therefore

$$\left(\frac{\partial \ln c}{\partial \phi}\right)_\theta = -\frac{1}{A}\left(\frac{\partial \ln c}{\partial \theta}\right)_\phi \left(\frac{\partial \sigma_m}{\partial \ln c}\right)_\phi = -\frac{1}{A}\left(\frac{\partial \sigma_m}{\partial \theta}\right)_\phi \qquad 4.91$$

But, according to equation 4.86

$$\frac{d\ln B}{d\phi} = \frac{1}{A}\left(\frac{\partial \sigma_m}{\partial \theta}\right)_\phi \qquad 4.92$$

Integration yields

$$\sigma_m = A\left(\frac{d\ln B}{d\phi}\right)\theta + \sigma_0 \qquad 4.93$$

σ_0 is the charge density on the electrode where $\theta = 0$. Let us denote the charge density on the electrode where $\theta = 1$, σ_1.

$$\sigma_1 = A\left(\frac{d\ln B}{d\phi}\right) + \sigma_0 \qquad 4.94$$

Assume now that the interphase where adsorption takes place can be regarded as two joint interphases, one where $\theta = 0$ and the other where $\theta = 1$. We can thus write another expression for σ_m

$$\sigma_m = \sigma_0\,(1-\theta) + \sigma_1\,\theta \qquad 4.95$$

But if the above assumption is correct the interphase can also be regarded as two capacitors in parallel

$$C_0 = \frac{d\sigma_0}{d\phi}$$ 4.96a

$$C_1 = \frac{d\sigma_1}{d\phi}$$ 4.96b

If we assume C_0 and C_1 to be independent of potential we get by integration

$$\sigma_0 = C_0(\phi - \phi_m)$$ 4.97a

$$\sigma_1 = C_1(\phi - \phi_m)$$ 4.97b

ϕ_m is the integration constant. Combination of equations 4.95 and 4.97 yields

$$\sigma = C_0(1-\theta)(\phi - \phi_m) + C_1\theta(\phi - \phi_m)$$ 4.98

$$\left(\frac{\partial\sigma}{\partial\theta}\right)_\phi = (C_1 - C_0)(\phi - \phi_m)$$ 4.99

and combination with equation 4.92 gives

$$\frac{d\ln B}{d\phi} = \frac{1}{A}(C_1 - C_0)(\phi - \phi_m)$$ 4.100

integration of which yields the desired expression

$$\ln B - \ln B_m = -J(\phi - \phi_m)^2$$ 4.101

$$J = -(C_1 - C_0)/2A$$

or

$$B = B_m \exp - J(\phi - \phi_m)^2$$ 4.102

Thus the free energy of adsorption depends on the square of the difference between the electrode potential and the potential of maximum adsorption, ϕ_m. ϕ_m is not identical with the potential of zero charge, but it is usually very near it.

It may be recalled that an equivalent dependence of the equilibrium constant of adsorption on potential for specifically adsorbed ions was not derived, because it depends not only on the electrode potential, but also on the potential at the o.h.p. The latter depends on the composition and concentration of the solution. However, an accepted working hypothesis is that the standard free energy of adsorption of ions varies linearly with the electrode charge.

The last problem to be dealt with in this section is that of assignment of adsorption isotherm from experimental data. In Chapter 3, Section A, three alternative adsorption isotherms were presented. In the literature on adsorption there are many more forms of isotherms. It seems that the task of assigning proper adsorption isotherms would be a simple one—surface excesses are easily obtained from electrocapillary measurements. However, whenever there is adsorption of no more than a monolayer on the electrode, the Γ vs. c relationship is bound to rise from zero to a maximum stationary value for Γ_m. The exact form of this rise, the adsorption isotherm, may be difficult to determine if Γ values are not very accurate. Experimental Γ values are often not accurate enough for the unambiguous assignment of a particular isotherm. Other methods for assigning isotherms from capacity data and from surface pressure data have been developed, but are outside the scope of this treatment.

In this section we saw that adsorption of neutral molecules behaves very differently from that of charged species. The greatest difference being the quadratic dependence of the free energy of adsorption on potential for neutral molecules, while for ions no simple relationship between B and potential exists.

D. The influence of the interphase on electrode kinetics

In Section A of this chapter four ways by which electrode kinetics are influenced by the properties of the interphase were discussed. These are: (1) the potential at the reaction site is not identical with the potential of the electrode; (2) molecules are polarized by the strong electric field of the interphase region; (3) the concentration of ions at the reaction site is influenced by the potential and (4) the availability of suitable sites may limit the reaction rate. In this section we shall try to illustrate these by looking at some examples of reactions which were studied specifically in order to evaluate the influence of the interphase on electrode reactions. It is hoped that these examples will also show the kind of phenomena to be observed while studying electrode reactions.

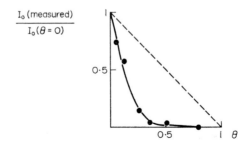

FIG. 38. Influence of the adsorption of cyclohexanol on the reduction of zinc on mercury.

The simplest way to illustrate the influence of the availability of suitable sites on the electrode on an electrode reaction is to add to the solution a known amount of material which adsorbs to a known extent and study the reaction with and without the presence of this known surface active substance. The inhibiting effect of the adsorbed film is shown in Fig. 38. This curve shows the effect of adsorption of cyclohexanol on the reduction of zinc at a zinc amalgam electrode. One clearly sees that when θ reaches $\theta = 0{\cdot}5$, the inhibiting effect of the cyclohexanol is total, i.e. no current is observed. On the other hand, adsorbed species can provide a way for an alternative path for the mechanism of the reaction studied and may, as a result, act as a catalyst rather than as an inhibitor. The catalytic effect of diphenylamine on the reduction of water is shown in Fig. 39: hydrogen is

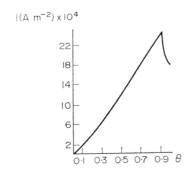

FIG. 39. Influence of the adsorption of diphenylamine on catalytic evolution of hydrogen on mercury.

evolved on mercury only at potentials more negative than 2 volts vs. the normal calomel electrode. However, on the introduction of diphenylamine to a solution of $0{\cdot}2\,M\,HCl$, one observes that hydrogen is being evolved at potentials round $-0{\cdot}8$V. The height of this wave, which is appropriately called "catalytic wave" depends on the surface concentration of the catalyst in the way illustrated in the figure.

The effect of polarization of the reactant in the electric field of the interphase has not been studied as such, since it would be very difficult to separate this effect from the others to be discussed presently. The two remaining factors in the influence of the interphase properties on electrode processes will be considered simultaneously.

The concentration of ions at a point of potential ϕ is

$$c = c_0 \exp -zF\phi/RT \qquad\qquad 4.103$$

where c_0 is the concentration of either the reductant or the oxidant in the absence of electric field. Introducing the Butler–Volmer equation gives

$$\frac{i}{nF} = \vec{k}'[R]_0 \exp - \left(\frac{z_R F\phi}{RT} \right) \exp \left(\beta \frac{nF}{RT} E' \right)$$

$$- \overleftarrow{k}'[Ox]_0 \exp - \left(\frac{z_{Ox} F\phi}{RT} \right) \exp - \left[(1-\beta) \frac{nF}{RT} E' \right] \qquad 4.104$$

Note that z is the valence of the ion and n is the number of electrons passed in a unit reaction.

$$z_{Ox} = z_R + n \qquad 4.105$$

The charge of the oxidized species is always more positive than that of the reduced species; the z values are taken with sign and n is always positive. E' in equation 4.104 is the rational potential $(E - E_{pzc})$; therefore the reaction rate constants differ from those used previously. But, as was said before, the potential which "drives" the electrode reaction, does not include that between the outer Helmholtz plane and the solution. The correct potential to be used in equation 4.104 is $E' - \phi_0$. Thus equation 4.104 is correctly written as

$$\frac{i}{nF} = \vec{k}'[R]_0 \exp - \left(\frac{z_R F\phi_0}{RT} \right) \exp \left[\beta \frac{nF}{RT} (E' - \phi_0) \right]$$

$$- \overleftarrow{k}'[Ox]_0 \exp - \left[\frac{(z_R + n)F\phi_0}{RT} \right] \exp - \left[(1-\beta) \frac{nF}{RT} (E' - \phi_0) \right] \qquad 4.106$$

Introducing the exchange current density i_0 and clearing brackets yields

$$\frac{i}{i_0} = \left\{ \exp - \frac{F}{RT} (z_R + \beta n)[\phi_0 - \phi(\eta = 0)] \right\}$$

$$\times \left\{ \exp \left(\beta \frac{nF}{RT} \eta \right) - \exp - \left[(1-\beta) \frac{nF}{RT} \eta \right] \right\} \qquad 4.107$$

This equation is known as the Frumkin equation, expressing the Frumkin theory. It shows the corrections which have to be made to a current-potential curve because of the presence of the double layer. However, it has two drawbacks. The first is that the site of reaction is identified with the plane of closest approach. In solutions of several components there are several such planes and the reaction site need not necessarily coincide with any of them. The second drawback is that this theory was developed for a simple, one step reaction and cannot be applied generally. However, equation 4.107 does provide us with some understanding of the influence of the value of ϕ_0

on the electrode current: in concentrated solutions ϕ_0 is very small and should not, therefore, influence the electrode reaction while in dilute solutions it will. This is shown in Fig. 40 for the reduction of chromate in solutions of alkali metal hydroxide. In dilute solutions $E_{\frac{1}{2}}$ is much more negative than in concentrated ones, indicating the difficulty of reduction in dilute solutions.

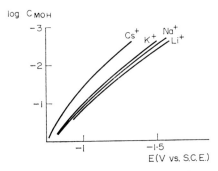

FIG. 40. Influence of supporting electrolyte cation on the half wave potential of chromate reduction ($2 \times 10^{-4}\,M$ chromate).

Equation 4.107 also shows a possibility of enhancing the electrode current by making $[z_R + (1 - \beta)n]\phi_0$ negative. This simply means that ions are attracted to electrodes of the opposite sign. This is well illustrated in Fig. 40 for reduction of the anion chromate and in Fig. 41 for the oxidation of Eu(II). The former process takes place on a negatively charged electrode and is, therefore, rather difficult. However when ϕ_0 is made more positive (at a given concentration of supporting electrolyte) by using cations which adsorb on the electrode, such as Cs^+, the reduction is facilitated. Figure 41

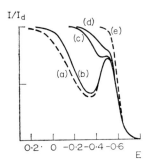

FIG. 41. The influence of the supporting electrolyte on the oxidation of Eu^{2+}; (a) $2 \times 10^{-3}\,M\,Eu^{2+}$ and $3 \times 10^{-3}\,M\,HClO_4$, (b) $+5 \times 10^{-4}\,M\,NaCl$, (c) $+5 \times 10^{-4}\,M$ NaBr, (d) $+5 \times 10^{-4}\,M\,NaI$, (e) a reversible process.

shows current-voltage curves for oxidation of Eu(II). The lowest curve was taken with a low concentration $(3 \times 10^{-3} M)$ of $HClO_4$. The other curves were taken in similar solutions with the addition of $5 \times 10^{-4} M$ of the sodium salts of the indicated anions. The uppermost curve is a calculated curve for

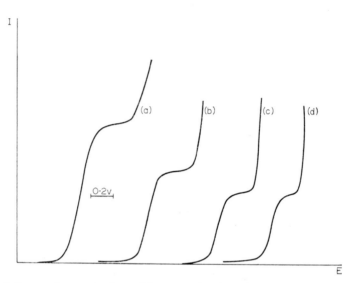

FIG. 42. Influence of concentrations of the supporting electrolyte on the reduction of HCl; (a) $10^{-4} M$ KCl, (b) $10^{-3} M$ KCl, (c) $10^{-2} M$ KCl, (d) $0 \cdot 1 M$ KCl.

the reversible process. It is seen that Eu(II) starts its oxidation on a negatively charged electrode, but as we scan in the positive direction, we pass through the p.z.c. to a positively charged electrode, where Eu^{2+} cations are repelled. This results in a decrease in current which increases again only at a very large overpotential. However, when specifically adsorbable anions are added, this effect diminishes until it almost disappears. Note the small concentration of supporting electrolytes: in much higher concentration this effect is not observed because ϕ_0 is too small. Figure 4.42 shows the reduction of HCl at a negatively charged electrode for various concentrations of KCl: 10^{-4}, 10^{-3}, 10^{-2}, $10^{-1} M$. That the hydronium cation is attracted to the negatively charged electrode when $\phi_{\text{o.h.p.}}$ is not too small is obvious from the very large limiting current observed. When the concentration of KCl is more than $0 \cdot 01 M$, however, $\phi_{\text{o.h.p.}}$ becomes effectively zero and the limiting current becomes equal to the diffusion current.

This chapter dealt with the structure and properties of the electrode solution interphase. It showed the difficulties encountered when trying to understand the interphase and the behaviour of adsorbed species on it. The fact that it dealt only with the interphase between mercury and aqueous solutions and that explanations for a number of phenomena was not given, shows how much there is still to be done in order to be able to predict the behaviour of a given electrode reaction at a given interphase.

Appendix
Derivation of Equations 4.75 and 4.76

We show the derivation of an equation for σ'_+. The derivation of an expression for σ'_- is done by a similar procedure.

$$\sigma'_+ = \frac{z_+ F n^0}{N} \int_{o.h.p.}^{\infty} \left[\exp - \left(\frac{z_+ e\phi}{kT} \right) - 1 \right] dx \qquad (4.73)$$

Consider the transformation

$$e^{-x} - 1 = (e^{-2x} - 2e^{-x} + 1)^{\frac{1}{2}} = [e^{-x}(e^{-x} - 2 + e^x)]^{\frac{1}{2}} \qquad A.1$$

so that

$$\sigma'_+ = \frac{z_+ F n^0}{N} \int_{o.h.p.}^{\infty} \left\{ \exp - \left(\frac{z_+ e\phi}{kT} \right) \left[\exp - \left(\frac{z_+ e\phi}{kT} \right) - 2 + \exp \left(\frac{z_+ e\phi}{kT} \right) \right] \right\}^{\frac{1}{2}} dx \quad A.2$$

Refer now to the arguments that led to equation 4.18. It is recalled that the Gauss surface was constructed to include all of the charge in the diffuse double-layer. If, however, we draw the surface in a way similar to that of Section B, 2, but so that it only includes part of the charge we find that

$$\sigma = K\varepsilon^0 \frac{d\phi}{dx} \qquad A.3$$

where σ is a function of x. Equation 4.19 then takes the form

$$\sigma(x) = -(8kTK\varepsilon^0 n^0)^{\frac{1}{2}} \sinh \frac{ze\phi}{2kT}. \qquad A.4$$

Consider the transformation

$$2 \sinh x = e^x - e^{-x} = (e^{2x} - 2 + e^{-2x})^{\frac{1}{2}} \qquad A.5$$

which, when applied to equation A.4 means

$$\sigma(x) = -(2kTK\varepsilon^0 n^0)^{\frac{1}{2}} \left[\exp \left(\frac{ze\phi}{kT} \right) - 2 + \exp - \left(\frac{ze\phi}{kT} \right) \right]^{\frac{1}{2}}. \qquad A.6$$

Combination of the above with equation A.2 yields

$$\sigma'_+ = -\frac{z_+ F n^{0\frac{1}{2}}}{N(2kTK\varepsilon^0)^{\frac{1}{2}}} \int_{o.h.p.}^{\infty} \sigma(x) \exp - \frac{z_+ e\phi}{2kT} dx. \qquad A.7$$

But from equation A.3

$$\sigma'_+ = -\frac{z_+ F}{N}\left(\frac{n^0 K \varepsilon^0}{2kT}\right)^{\frac{1}{2}} \int_{\text{o.h.p.}}^{\infty} \exp - \frac{z_+ e\phi}{2kT}\, d\phi. \qquad \text{A.8}$$

Integration yields

$$\sigma'_+ = (2kT K \varepsilon^0 n^0)^{\frac{1}{2}}\left[\exp - \left(\frac{z_+ e\phi_0}{2kT}\right) - 1\right]. \qquad \begin{array}{l}\text{A.9}\\(4.75)\end{array}$$

5. Techniques of Measurement

Up to this point we have discussed various theoretical aspects of electrodics. We have examined the mechanisms which may be followed by an electrode reaction and have looked into that all important region, the interphase. This chapter, as its title implies, will deal with experimental techniques, i.e. it will attempt to answer the ever present question of how all this knowledge is assembled. We shall start with a general discussion of the various functions that are measured and divide the area of electrodic experimentation into three main groups. We shall proceed to deal with various experimental practices which are general to all techniques and discuss each of the three groups separately, confining ourselves mostly to reactions where both electron transfer and diffusion together are the rate determining steps. The chapter will close with a section on the newest techniques in the study of electrode processes, the spectroscopic techniques.

A. What is measured?

In the previous chapters we met various functions: potential, current, surface tension, capacity and concentration. Time has not been mentioned explicitly but, as in ordinary kinetics, it is one of the variables which are of interest.

The techniques to be described below, all measure or control one or more of these variables. Time is always a measured, but independent, variable. Current, potential and concentration are examples of variables which can be controlled in some methods or measured as dependent variables in others. Every technique has its own set of variables some of which are measured, some controlled and some held constant.

The object of our measurements is to determine the values of the several-rate constants of the electrode reaction: any electrode reaction may involve several steps, each step will have a rate constant and will be influenced by diffusion, adsorption and other parameters in a way different from the other steps. The explicit form of the interdependence of these parameters, concentration, electrode potential and reaction rate are of great interest to the electrochemist.

Most techniques used in the study of electrode reactions and the electrode-solution interphase can be divided into three groups: (a) equilibrium measurements (b) steady state measurement and (c) transient measurement. The first group includes classical potentiometry, measurements of properties of interphases and of impedance of cells. Since it is the equilibrium properties of the system under study which are of interest, the theories used are thermodynamic in approach. The second group includes classical polarography, voltammetry in stirred solutions and a.c. polarography. Voltammetry in quiescent solutions is not strictly a steady state technique for reasons which will become apparent later. The term "steady state" implies that the function measured does not change with time, but the system is not at equilibrium. Thermodynamic methods do not apply here, the theories used for the interpretation of steady state measurements are kinetic. The third group includes all techniques where the independent, controlled variable is changing rapidly with time. The dependent variables are usually measured as functions of time from which the desired kinetic parameters are calculated. This group also includes cyclic voltammetry where time is not measured as an explicit variable. It is, however, measured as a third variable and influences markedly the current potential curves obtained.

B. Cells, electrodes and equipment

Solutions in electrochemical cells are generally composed of the reactant, any other inert material that may be present for various reasons and the supporting electrolyte. The function of the supporting electrolyte may now, after having discussed the properties of the interphase, be better understood. First it provides for the conductivity of the solution. Since reactants are often present in millimolar concentrations, an electrolyte must be added to minimize the IR drop in the solution. Second it eliminates migration in the electric field as one of the modes of mass transport of the reactant ions; this can be looked at by remembering the formula for the transport number (equation 2.16)—the greater the concentration of a particular type of ion, the greater its transport number. Thus the addition of a high concentration of electrolyte makes the transport number of the reactant negligible, i.e. it does not move under the influence of the electric field. The third way in which the supporting electrolyte acts is by almost completely eliminating the existence of a diffuse double layer, all of the potential drop takes place between the electrode and the outer Helmholtz plane (which is probably the reaction site) thus minimizing the effects of the interphase structure on the electrode process under consideration. It is customary to make the concentration of the supporting electrolyte fifty to one hundred times greater than that of the reactant.

1. *Cells*

All cell-electrode arrangements can be divided into two groups. The first group includes the two electrode arrangement and the second, the three electrode arrangement. Every cell must have an anode and a cathode and each one may be the working electrode. If we are interested in the oxidation the anode is the working electrode, if we study reduction, the cathode is the working electrode. The potential of the working electrode is measured or controlled, hence the need for a reference electrode. In the two electrode arrangements the reference electrode also serves as a counter electrode. This arrangement is suitable only for cases where the current passed through the

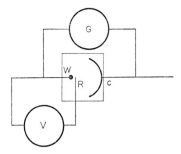

FIG. 43. Schematic presentation of a three electrode system; *W*: working electrode, *R*: reference electrode, *c:* counter electrode, *V*: voltmeter of a very high resistance, *G*: galvanometer or current measuring recorder.

cell is very small, so small that the reference-counter electrode can be regarded as ideally non-polarizable (i.e. of constant potential). In other cases, i.e. when the current passed through the cell is too large, a three electrode system must be used as shown in Fig. 43. Since the potential of the reference electrode is constant, any change of cell potential may be regarded as a change in the potential of the working electrode.

The need to separate the functions of counter and reference electrodes arises when (*a*) the current passed is too high and may polarize a counter-reference electrode, (*b*) when the *IR* drop in the solution is too large either because of the solution resistance or because of the high current.

The first situation arises when the current-potential curve of the reference electrode deviates from the ideal one for a non-polarizable electrode by more than the error in voltage measurement. For example, in classical polarography as used in routine analytical applications the potential is seldom measured more accurately than 10 m̃V. A reference electrode whose potential deviates less than that under the conditions of measurement could be very well used as a counter-reference electrode in a two-electrode arrangement. If, however one must use a reference electrode whose potential deviates considerably from that of equilibrium, e.g. in a non-aqueous medium, then a three electrode arrangement is essential. The most popular two electrode cell is the H-cell. This consists of two compartments, one containing the working electrode and the test solution and the other containing the reference electrode half-cell. The two parts are usually joined by a fritted glass disk. The three electrode cells can be divided into two groups, one where the counter electrode is placed in the same compartment as the working electrode and the other where the two electrode compartments are separated by a partition that allows ionic movement, but not mixing of the solutions. The first type of cell may be used when the reaction products of the working and the counter electrodes do not react with each other; if they do, it is advantageous to use the second type of cell. A full description of all the cells reported in the literature for various applications will fill many pages and is quite pointless. Given a particular reaction studied in a particular way, a proper design for a cell will usually suggest itself.

In the three electrode arrangement, we usually, but not always, place the tip of the reference electrode very near the surface of the working electrode. This is done in order to minimize the electric field, which is generated by the IR drop, between the working and reference electrodes and hence get a true reading of potential. The equivalent circuit of a cell is shown in Fig. 44 in

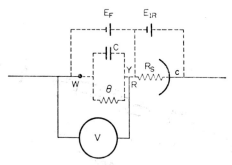

FIG. 44. The placement of the reference electrode in a three electrode arrangement.

dotted lines, R_s acts toward the reference electrode as the potentiometer resistance toward the sliding contact. If the sliding contact is close to Y, the potential registered by the voltmeter V would be the faradaic potential E_F plus the IR drop E_{IR}. If the sliding contact is in the position shown, V will read only E_F.

2. Electrodes

Let us now discuss some of the experimental principles of the construction of working, counter and reference electrodes. The working electrode must be made of a material that will not react with the solvent or any component of the solution over as wide a potential range as possible. Figure 45 shows two current potential curves taken with the d.m.e. and platinum electrodes in aqueous KCl solutions showing the potentials beyond which these cannot be used as working electrodes. The potential limit beyond which an electrode cannot be used may be defined in several ways. One useful way is to extrapolate the horizontal portion of the current-potential curve (called the "residual current") and measure the potential where the current is 5 μA higher than the extrapolated value. Thus the potential limits of mercury in KCl are 0 and $-2 \cdot 1$ volts, of platinum $+0 \cdot 6$ and $-0 \cdot 6$ volts (from Fig. 45)

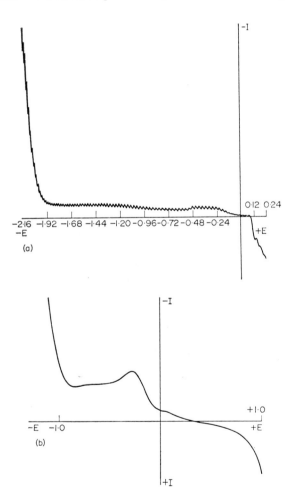

FIG. 45. Current potential curves in $0 \cdot 1$ M KCl solution; (a) The d.m.e., (b) Pt electrode.

and of graphite in KCl $+1\cdot3$ and $-1\cdot6$ volts. The sharp rise in current may be due to various kinds of processes. For example, the anodic rise in the case of graphite is due to the oxidation of water, in the case of platinum to the formation of the platinum-chloride complex on the surface of the electrode and in the case of mercury, to the oxidation of the mercury metal itself. The rise of cathodic current is due, in the cases of platinum and graphite, to evolution of hydrogen but in the case of mercury, to the reduction of K^+ and the consequent formation of potassium amalgam. Working electrodes can be large or small, depending on the particular application. One usually prefers smooth, rather than rough, electrodes because mass transport processes to a smooth electrode are better defined.

Counter electrodes are usually made of metal foils which react electrolytically with one or more of the components of the solution. It is very important to make sure that the solution contains such a reactive component because if it does not, current will not pass through the cell. Sometimes there must be added to the counter electrode compartment a component whose sole function is to be the electroactive material for the reaction at that electrode.

The potential of reference electrodes should be constant, established quickly and not vary much if a few microamperes are passed through the electrode. In other words, the electrode should have a minimal resistance. This resistance should be independent of the direction of current flow and should

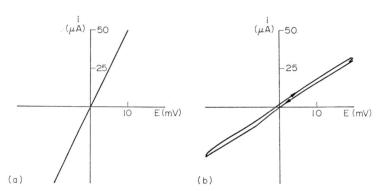

FIG. 46. Current potential curves for reference electrodes: (*a*) Good, (*b*) Bad.

be constant. This implies that the current–potential curve for a reference electrode should look like that in Fig. 46(a) and not like the one in Fig. 46(b). The equation describing this desired current–potential curve is the linear Butler–Volmer equation 3.71.

$$i = i_0 nF\eta \qquad\qquad 3.71$$

The resistance of the electrode is then

$$R = \frac{\eta}{i} = \frac{1}{nFi_0} \qquad 5.1$$

which means that the reference electrode should have a high exchange current-density. Recall the expression for the exchange current density

$$i_0 = nFk_s[R]^{1-\beta}[Ox]^{\beta} \qquad 3.68$$

This equation shows that in order to construct a reference electrode with a low resistance, the electrode reaction should be rapid (high k_s) and the concentrations of the materials making up the electrode should be high. The numbers of metal-ion systems suitable for the construction of reference electrodes is limited to those where the reaction rate constant k_s is high. However, the suitable metal-cation systems can be used with many anions and, usually, in several solvents, making the number of complete reference electrodes rather large.

The most useful electrode in studies of aqueous systems is the saturated calomel electrode, a schematic drawing of which is given in Fig. 47. It is made up of a pool of mercury covered with a paste of wet calomel and

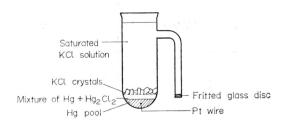

FIG. 47. A schematic drawing of a saturated calomel reference electrode.

TABLE IX. *Potentials of several aqueous reference electrodes (volts against the standard hydrogen electrode at 25°C)*

Electrode symbol	Electrode reaction	E or E⁰	
Hg/Hg$_2$Cl$_2$/KCl (sat)	$2Hg + 2Cl^- = Hg_2Cl_2 + 2e$		0·2412
Hg/Hg$_2$Cl$_2$/KCl (1 M)	$2Hg + 2Cl^- = Hg_2Cl_2 + 2e$		0·2801
Ag/AgCl/Cl$^-$	$Ag + Cl^- = AgCl + e$	$E^0 =$	0·222
Ag/AgBr/Br$^-$	$Ag + Br^- = AgBr + e$	$E^0 =$	0·095
Ag/AgI/I$^-$	$Ag + I^- = AgI + e$	$E^0 =$	$-0·151$
Hg/Hg$_2$SO$_4$/SO$_4$$^{2-}$	$2Hg + SO_4^{2-} = Hg_2SO_4 + 2e$	$E^0 =$	0·6151

mercury, in contact with a saturated solution of KCl containing some solid KCl. Also very useful are the various silver halide electrodes and the mercury-mercurous sulphate electrode. Table IX gives potential values of these commonly used aqueous electrodes as measured against the standard hydrogen electrode.

Reference electrodes in many non-aqueous solvents are known. Their construction follows the same principles as for aqueous reference electrodes—choose a rapid reaction and make sure that the concentrations of the reactants and products are high. Many non-aqueous reference electrodes use reactions based on mercury and silver, but lead, thalium and other metals have also been used.

3. Electronic Equipment

The electronic equipment used in the study of electrode processes may vary considerably in degree of sophistication and in price. However, it can conveniently be described by referring to the simple polarographic circuit which is given in Fig. 48. This basic circuit is made up of a variable voltage

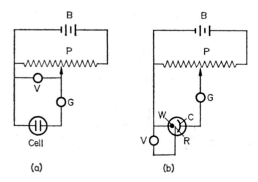

FIG. 48. Schematic presentation of a polarographic circuit; (a) A two electrode arrangement; (b) A three electrode arrangement.

source, a cell and a current measuring instrument. In a two electrode arrangement the voltage across the cell is taken as the potential of the working electrode. In a three electrode arrangement the voltage of the source is applied across the anode and cathode and the working electrode potential is measured between the latter and a reference electrode using a suitable instrument. Let us now consider voltage sources, current measuring devices and voltage measuring equipment.

The simplest variable voltage source is a battery connected across a potentiometer and adjusted manually. The batteries can be replaced by a mains operated rectifier and the hand operated potentiometer may be replaced by a synchronous motor to drive the potentiometer and so provide a linear change of voltage with time. Faster scan rates than are practical with a motor driven potentiometer can be produced by using operational amplifiers. Often one may not want a linear scan but a step function of potential, for which square wave generators are used.

The simplest current measuring instrument is the galvanometer, but its sensitivity is limited to small currents, the current cannot be conveniently recorded and its operation is slow. If the galvanometer is replaced by a current measuring resistor, the potential drop across the latter is proportional to the current passed. Thus, the current measuring problem is converted to a voltage measuring problem. Here potentiometric recorders are very useful when the rate of change of voltage is much lower than the response time of the recorder. For faster changes, oscilloscopes are used. The recorder or oscilloscope must be carefully selected so that the response time of the measuring instrument (the "rise time") will be much smaller than the duration of the measurement. This is particularly important in oscilloscopes used with very high scan rates. Sometimes very sensitive oscilloscopes do not have the necessary rise time and a choice between sensitivity and speed must be made. Another important point to remember is that the current passed in the recorder or oscilloscope must be a very small fraction of the current being measured. For example, if the current measured is 10^{-5} A and the sensitivity of the recorder is 10^{-2} V, the resistor through which the current has to pass is 1000 ohms. If the current used by the recorder is not to exceed one thousandth of the current passed through the measuring resistor, the resistance of the recorder should be greater than 10^6 ohm. Thus there are two factors to be considered when using a recorder or an oscilloscope, one is the response time and the other is the impedance of the instrument.

The graph drawn by a strip-chart recorder is invariably that of voltage against time. Thus any quantity that can be converted to voltage, such as current, temperature, etc. can be recorded against time. In the experimental set-up for classical polarography, the voltage applied to the cell is a linear function of time, thus the graph from a strip-chart recorder can be interpreted as a current potential curve. In the three electrode arrangement, the potential may not strictly be a linear function of time, making the use of strip-chart recorders rather inconvenient. In order to record current voltage curves directly, one may use $x-y$ plotters, where one voltage drop (across the standard resistor) is fed to the y axis and the other (between the working and reference electrodes) is fed to the x axis. Many oscilloscopes also have $x-y$ operation and can be used for fast recordings. The same criteria of response time and impedance also apply here. Since we do not want to pass large currents through the reference electrode, the measuring instrument, i.e.

recorder or oscilloscope, must have a high input impedance. For example, if we want to pass through the reference electrode no more than one micro-ampere, the resistance of the recorder must be one megohm per volt. Instruments of this, and higher, impedances are available, but have to be specifically looked for.

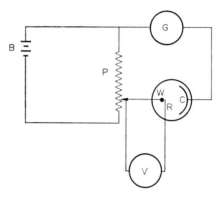

FIG. 49. A manual potentiostat (whenever the reading on the voltmeter V differs from the wanted one, the potentiometer P is adjusted until the correct voltage is obtained.

Often, when using three electrode arrangements we are interested in controlling the rate of change of potential difference not between the cathode and the anode, but between the working and reference electrodes. In other words, we want to adjust the potential difference between the latter two electrodes to a desired value and apply that voltage between the anode and cathode as needed. This can be done manually as shown in Fig. 49, or auto-matically using the instrument known as a potentiostat. A schematic diagram of a potenstiostat is given in Fig. 50 and a detailed explanation is

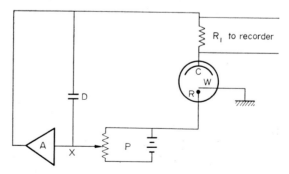

FIG. 50. A schematic diagram of an automatic potentiostat; A: operational amplifier, D: a large capacitor, R_1 a measuring resistor.

given in the Appendix at the end of this chapter. The working electrode of the cell is grounded, and the potentiometer P is set at the desired potential between the reference and working electrodes. The operational amplifier keeps the point x at ground potential by passing enough current through the cell thus maintaining the reference electrode potential at a suitable level vs. ground. The current passing through the reference electrode is very small because the input impedance of the amplifier is very high. The potential of P need not be constant, it can be modulated using waveform generators or motor driven potentiometers to suit the need. However, there is a limit to the speed at which the reference potential can be changed. There is a time lag between the setting of P and the establishment of ground potential at x, this lag is often called the "rise time" of the potentiostat and should be very much smaller than the duration of the measurement.

As was said before, three electrode arrangements and, therefore, potentio-stats, are used when the resistance of the solution is high. Thus a potentiostat should be capable of providing that voltage between the cathode and anode as needed by the current passed and the resistance of the solution. For example, on a macroscale electrolysis in non-aqueous solutions, the current passed may be several milliamperes and the resistance of the solution may be several tens of thousands of ohms; the potential difference between the cathode and anode thus should be of the order of tens of volts. The number of commercially available potentiostats is increasing rapidly and shorter rise time and higher voltage across the cell should become available as more potentiostats are made.

The current potential curve of oxygen on the d.m.e. is shown in Fig. 51. The curve consists of two waves, one very close to the potential of the s.c.e.

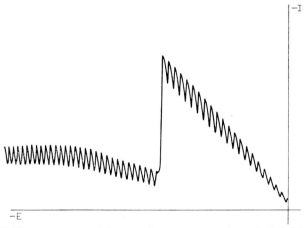

FIG. 51. The current potential curve of oxygen on the d.m.e. in $0·001\ M$ KCl.

and the other about one volt negative to it. When studying reductions on the d.m.e., the constant presence of this oxygen wave is usually very undesirable. The best way to eliminate it is to purge the oxygen out of the solution using some other gas, the commonest being nitrogen. Argon and helium are also used for special purposes. Bottled nitrogen gas is usually quite pure and may be used in routine analytical and other applications without further purification. However, when the solutions studied are very dilute or when an organic solvent is used, the trace quantities of oxygen that this nitrogen contains must be removed. A good practice is to bubble the nitrogen into a solution of the solvent studied and the supporting electrolyte prior to passing it into the test solution in order to saturate it with the solvent vapour; if this is not done the continuous passage of gas through the test solution may evaporate part of the solvent, increasing the concentrations of the solutes. The reduction wave of oxygen is not peculiar to the d.m.e. and appears whenever the electrode reaction takes place at potentials negative to that of the s.c.e.

C. Measurements at equilibrium

Measurements at equilibrium can be divided into three separate categories, potentiometric measurements, measurements of the properties of interphases and impedance measurements. The first group will not be discussed here since the various theoretical and experimental aspects of potentiometry are very well covered by texts on solution electro-chemistry and on analytical chemistry.

1. Measurement of the properties of interphases

This section will be divided into techniques applicable only to liquid electrodes, techniques used for both liquid and solid electrodes and special techniques and problems associated with solid electrodes.

The charge on an electrode, q_m, can be calculated in two ways. It can be calculated directly from Gibbs adsorption isotherm (equation 4.58) or it can be determined from data on the capacity of the double layer C.

$$q_m = \int_{E_{pzc}}^{E} C \, dE \qquad \qquad 5.2$$

Both the surface tension and capacity are measurable quantities for liquid electrodes, and parameters of the interphase, such as surface excesses, can be determined from either quantity.

Surface tension can be measured mainly by two methods, the first uses the capillary electrometer and the second measures the weight of falling drops.

The capillary electrometer measures the pressure (P) on the meniscus of mercury in the capillary

$$P = \frac{2\gamma \cos \theta}{r} \qquad\qquad 5.3$$

by balancing it against the height (h) of the mercury column which is supported by this meniscus

$$P = hg\rho \qquad\qquad 5.4$$

where θ is the contact angle between mercury and the capillary wall, r is the radius of curvature of the meniscus, γ is the surface tension of the mercury, g is the acceleration due to gravity and ρ is the density of mercury. Thus

$$\gamma = \frac{hg\rho}{2 \cos \theta} \qquad\qquad 5.5$$

A schematic picture of the capillary electrometer is given in Fig. 52. The solution is contained in the cell C into which the capillary tip T is immersed. The mercury column above this tip is connected via a side arm to a mercury reservoir R, the height of which can be adjusted. Alternatively, the top of

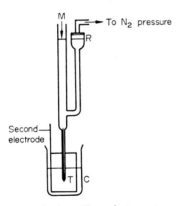

Fig. 52. A capillary electrometer.

the mercury reservoir may be connected to a nitrogen cylinder and the pressure on the capillary tip can be changed by changing the pressure at R. Commonly, the position of the mercury in the capillary is kept constant and the height of the mercury column is measured by measuring the position of the top meniscus M. In order to obtain reliable results from the capillary electrometer, θ should equal 180°, i.e. there should be a thin layer of solution between the mercury and capillary wall. In very dilute solutions this thin layer may break, making data obtained by this method unreliable.

It is rather difficult to determine the radius of curvature of the meniscus. Therefore equation 5.5 is simply written

$$\gamma = kh \qquad\qquad 5.6$$

and k is determined experimentally by using the known value of γ in acidified dilute Na_2SO_4 solution at the p.z.c.

$$\gamma_{max} = 0{\cdot}4267 - 0{\cdot}00017(t - 18)\,\tau\,m^{-2} \qquad\qquad 5.7$$

t being the temperature in degrees centigrade.

The second method to be discussed here is the determination of surface tension by measuring the weight of falling mercury drops. To a first approximation the weight of the drop (corrected for the weight of solution displaced by it) is proportional to the surface forces retaining it

$$\frac{\rho_{Hg} - \rho_{soln}}{\rho_{Hg}} M = 2\pi r\gamma \qquad\qquad 5.8$$

M is the weight of the drop and r is the inner radius of the capillary. M is given by the rate of flow of mercury m, times the drop time t_d. Thus determination of the drop time should give information about the surface tension. However, equation 5.8 is only an approximation and the surface tension values derived from it are less accurate than those measured using the capillary electrometer. The method is still very useful for measuring differences in surface tension which are needed for the determination of adsorption on mercury and in fact the results obtained compare very well with those obtained by the capillary electrometer method.

The two methods described here are not the only ones used for the determination of surface tension. For descriptions of other methods the reader is referred to the specialist literature.

The second quantity measured when studying the interphase is the capacity of the double layer. It may be recalled from equation 2.8 that the current passed through a cell is

$$I = nF dM/dt + C dE/dt \qquad\qquad 2.8$$

When the electrode is ideally polarizable, there is no reaction at its surface and the first term on the right hand side of the equation is zero. Thus, when the resistance of the solution is negligible compared to the impedance of the double layer, the current passed is proportional to the capacity. This gives one way of measuring capacities; the rate of change of voltage may be programmed in any way desired and the current potential curve recorded. The capacity of the double layer will be given by integration of equation 2.8

$$E = E_0 + (I/C)\,t \qquad\qquad 5.9$$

This method is limited to a very concentrated solution where the resistance of the solution is very low. For other solutions, the current is given by the following equation which is derived by considering the capacity and resistance in series

$$E = RI + (1/C) \int_0^t I dt \qquad 5.10$$

or

$$\frac{dE}{dt} = R \frac{dI}{dt} + \frac{I}{C} \qquad 5.11$$

There are two ways of eliminating the first term on the right hand side of equation 5.11, either by working with systems of negligible resistance or by working at constant current. Constant current generators are very common and by applying a current pulse and measuring the rate of change of voltage with time one arrives at values of the double layer capacity.

Of all the numerous methods devised for the measurement of double layer capacity, the most popular and most accurate has been the measurement of the impedance of the cell using an a.c. bridge. Figure 53 shows a schematic

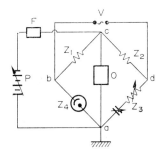

FIG. 53. A schematic diagram of an impedance bridge for measuring capacity of the double layer.

diagram of an impedance bridge where Z_4 represents the measuring cell. If $Z_1 = Z_2$ the bridge is balanced (i.e., the voltage between points a and c is zero) when $Z_3 = Z_4$. In other words, the oscilloscope O used as a null detector shows zero when the resistance at Z_3 is equal to the resistance of the cell and the capacity of Z_3 equals the capacity of the cell. If the a.c. voltage source V has frequency such that a period of one oscillation is much larger than the time of formation of the double layer, the measured capacity should not be frequency dependent. In experimental bridges, however, such a frequency dependence may be observed and usually indicates faults in the design of the bridge. Let us now discuss every component of Fig. 53 in detail and see what happens when the frequency of measurement changes. The impedances Z_1 and Z_2 are usually identical resistances. The impedance Z_3 is made of standard resistance and capacitance boxes. When the frequency is low, these give adequate readings, but when the frequency is high, the small inductance in the resistance boxes has to be compensated by shunting a known small inductance across them.

The geometry of the cell Z_4 is very important for capacitance measurements. Ideally the electric field should be spherically symmetrical round the working electrode. In practice, however, good results are obtained if the working electrode is a small sphere (such as the d.m.e.) and the counter electrode is a large cylinder of metal foil placed so that the working electrode is in its centre.

The filter F is usually a very high inductance. However, its filtering efficiency declines at low frequencies. This problem is solved by using a very large resistance for the filter at low frequencies. Another point in the design of the potentiometer, P and F is that their output impedance should be as high as possible in order to ensure that the a.c. current flows through the bridge and not through the potentiometer. The potentiometer-filter assembly shown in Fig. 53 is connected between points a and c, rather than points a and b. This is another way of ensuring the separation between the a.c. current in the bridge and the d.c. voltage of the potentiometer; when the bridge is balanced the potential at c is ground potential and no current passes through O or through P and F. The filter F is kept to separate the two components (a.c. and d.c.) when the bridge is not balanced.

The a.c. source V is usually of rather small amplitude, the currents it generates in the bridge become very small as the frequency goes up. Obviously, it is much more difficult to balance small currents and therefore one introduces a transformer between V and the bridge whose output resistance is very small and whose sole function is to lower the impedance of the bridge circuit thus providing for higher currents. All components of the bridge are protected from stray electric fields in the laboratory by grounded screens.

The double layer is often studied in order to determine adsorption of organic or inorganic species on the electrode. Obviously, we do not want the interference of impurities which adsorb on the electrode, sometimes to a greater extent than the studied material. Thus, a very careful purification of the solution must be carried out before measurement. If the electrode is of constant area (i.e. a solid or hanging drop or mercury pool) impurity adsorption increases with time, thus making the interference more serious at longer times of measurement. When working with the d.m.e., the rate of flow of mercury and, hence, the rate of formation of clean surface, can be made much more rapid than the rate of adsorption of impurities; but since the area of the electrode changes all the time, the current changes too and the bridge can only be balanced at one instance in the life of the drop. If one measures the time passed from the fall of the last drop to the time when the bridge is balanced, one can calculate the area of the drop and, hence, the capacity per unit area. Measurement of this time of balance is usually done with the time base on the oscilloscope which can very conveniently be used as a clock. The time of fall of the former drop is marked by a sharp change in current which can be used to trigger the oscilloscope. The balance point is when the trace crosses the zero current line.

Let us now turn our attention to the study of the double layer at solid electrodes. The methods described previously for the determination of the surface tension do not apply to solid electrodes. Information on surface tension can be obtained by relating changes of parameters, such as friction hardness and creep, with potential to electrocapillary curves. Curves showing changes of these parameters with potential look like electrocapillary curves and the maxima usually correspond to the potential of zero charge. Potentials of zero charge can also be determined by measuring the potential of the metal as it is being dipped into a solution, which initially is the p.z.c. Later, because of various chemical reactions between the metal and the solution, the potential will change (for example, to the Nernst equilibrium potential).

Measurement of the capacity of the double layer with solid electrodes does not differ in principle from that with mercury. However, there are several complicating factors with solid electrodes. The state of the surface of an electrode depends very much on its history, therefore, in order to obtain meaningful data care must be taken to treat the electrodes in a standard way. The surface of solids is usually rough, the "roughness factor" being the ratio of the true area of the electrode to its apparent geometric area. If the roughness factor is large, the capacity of the double layer will depend very much on the current frequency where the measurements were made. This frequency dependence can be reduced by polishing the electrode or by surface fusion, but cannot be completely eliminated. Thus all double layer measurements have to be extrapolated to zero frequency, such extrapolation always being an error introducing procedure. Many solid electrodes form surface films by reacting with the solution, e.g. the formation of adsorbed hydrogen layer on platinum. If we want to measure the properties of the electrodes at equilibrium, we must know the behaviour of the particular electrode in the solvents used so that we know the potential regions where such films are formed.

A brief discussion of cell impedance was given in Section G, Chapter 3 and the equivalent circuit of an electrochemical cell was given in Fig. 18. The following section will be concerned with the way by which impedance measurements are made and interpreted to give data on the kinetics of the electrode process studied.

The cell used for impedance measurements is constructed very much in the same way as that used for capacity measurements (Fig. 54). The large counter

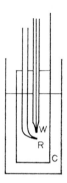

FIG. 54. A cell used for the measurement of capacity and impedance. W: the working electrode, R: the reference electrode; C: the counter electrode.

electrode has a very large capacity, therefore a small capacitive impedance. It also has a small resistance due partly to its large area and partly to the care that we take in order to provide a fast electron transfer reaction to occur at the counter electrode. Thus the value of Z_R in Fig. 18 is negligible compared to that of the impedance $\mathbf{Z}$ associated with the working electrode. The latter is usually broken up to the resistance of electron transfer, θ, which is given by equation 3.106 and to the impedance of the diffusion process, the Warburg impedance, W which is given as

$$W = \sigma\omega^{-\frac{1}{2}} - j\sigma\omega^{-\frac{1}{2}}$$

where

$$\sigma = \frac{RT}{n^2 F^2 2^{\frac{1}{2}}} \left(\frac{1}{[\mathrm{Ox}] D_{\mathrm{Ox}}^{\frac{1}{2}}} + \frac{1}{[R] D_R^{\frac{1}{2}}} \right) \qquad 5.12$$

j is the square root of -1 and ω is the angular frequency of the current or potential. When an expression for impedance contains an imaginary term (a term multiplied by j), it means that the circuit described also contains capacity and inductive components. In the case of electro-chemical cells one considers only capacity components of impedance.

The most accurate method for the determination of impedance is, again, by the use of a.c. bridges used for the measurement of capacity of the double layer. There is one difference, however. Since impedance measurements are usually made at the equilibrium potential, the need for the d.c. potential source does not arise. When working with the d.m.e. this component can be eliminated, but when working with solid electrodes, the d.c. source is needed in order to pretreat the working electrode in a way and for reasons explained in Section D.

The impedance of the cell is measured as a function of frequency for every equilibrium potential for as wide a frequency range as possible. The equilibrium potential is changed by changing the concentrations of the reactant and product in a known way (a reference electrode is included in the cell for measurement of these equilibrium potentials). For every value of equilibrium potential, the resistance measured at the bridge (at the Z_3 arm) R_3, is plotted against the value of $1/\omega C_3$, C_3 being the capacity measured. These two values are the real and imaginary components of the cell impedance, their behaviour as a function of frequency gives all the parameters needed to understand the kinetics of the electrode process. A short discussion of the theory of cell impedance will now follow.

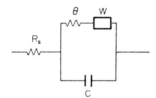

FIG. 55. Equivalent circuit for a cell used for impedance measurements; the electrode reaction is controlled by diffusion and electron transfer.

Figure 55 shows the equivalent circuit of a cell used for impedance measurements, Z_R is not shown and Z is given as the resistance of electron transfer θ and that of diffusion W. The impedance of this circuit is given by the following equation

$$Z = R_s + \cfrac{1}{j\omega C + \cfrac{1}{\theta + \sigma\omega^{-\frac{1}{2}} - j\sigma\omega^{-\frac{1}{2}}}} \qquad 5.13$$

which simplifies for extreme values of the frequency to the following: at low frequency

$$Z = Z' + jZ'' = R_s + \theta + \sigma\omega^{-\frac{1}{2}} - j(\sigma\omega^{-\frac{1}{2}} - 2\sigma^2 C) \qquad 5.14$$

at high frequency

$$Z = Z' + jZ'' = R_s + \frac{\theta}{1 + \omega^2 C^2 \theta^2} - j\frac{\omega C\theta^2}{1 + \omega^2 C^2 \theta^2} \qquad 5.15$$

or, for low frequency the real part of the impedance, Z' is

$$Z_1' = R_s + \theta + \sigma\omega^{-\frac{1}{2}} \qquad\qquad 5.14a$$

and the imaginary part, Z'' is

$$Z_1'' = -\sigma\omega^{-\frac{1}{2}} - 2\sigma^2\,C \qquad\qquad 5.14b$$

Similarly for high frequency

$$Z_h' = R_s + \frac{\theta}{1+\omega^2\,C^2\,\theta^2} \qquad\qquad 5.15a$$

$$Z_h'' = -\frac{\omega C\theta^2}{1+\omega^2 C^2\theta^2} \qquad\qquad 5.15b$$

Eliminating ω between 5.14a and 5.14b yields

$$Z_1'' = Z_1' + R_s + \theta - 2\sigma^2\,C \qquad\qquad 5.16$$

Thus, when one plots Z_1' against Z_1'', one obtains a straight line the slope of which is 45°, and the intercept (for $Z_1'' = 0$) is $-R_s - \theta + 2\sigma^2\,C$. If one eliminates ω between equations 5.15a and 5.15b one obtains

$$(Z_h' - R_s - \theta/2)^2 + (Z_h'')^2 = (\theta/2)^2 \qquad\qquad 5.17$$

The equation is that of a circle of radius $\theta/2$, the centre of which lies on the Z' axis at $Z_h' = R_s + \theta/2$. Figure 56 shows the curve obtained on plotting Z'

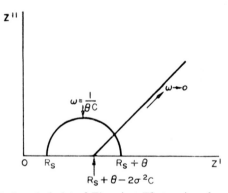

Fig. 56. A typical plot of Z' against Z'' at various frequencies.

against Z'' for various frequencies, but since only the positive values of the impedance are meaningful, Fig. 56 shows only a semi-circle.

At the apex of the circle $Z'' = \theta/2$. If one substitutes this value in equation 5.15b, and solves for ω, one gets $\omega = 1/\theta C$, i.e. the frequency where the apex of the circle is reached, gives (θ is known from the radius of the circle) the capacity of the double layer in the presence of an electrode reaction. R_s is obtained from the intersection of the semi-circle with the Z' axis. θ is proportional to $1/i_0$ (equation 3.106) and the diffusion parameter σ can be obtained from the intersection of the straight line portion of the Z' against

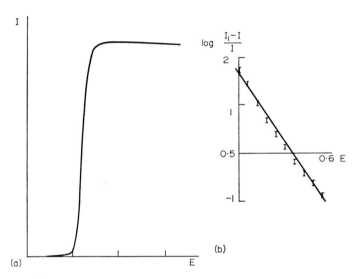

FIG. 57. Cd^{2+} reduction at the d.m.e.; (a) Current potential curve, (b) Log plot.

Z'' plot with the Z' axis. The transfer coefficient β can be determined in the usual way by studying the dependence of i_0 on the concentrations of reactant and product (equation 3.68).

The impedance method for determination of electrode kinetics was found to work for very fast electrode reactions. Measurements of i_0 up to 10^4 A m^{-2} (1 A cm^{-2}) are possible; higher values of i_0 can also be determined, but with less accuracy. The reason for the diminished accuracy for the extremely fast electrode reactions is that these have considerable diffusion component to the impedance even at fairly high frequencies, thus masking the form of the semi-circle. The form of the semi-circle can be distorted also if the capacity of the double layer, C, is very large. It is thus advisable to work with small indicating electrodes.

If the electrode reaction is preceded by a chemical reaction or if it includes adsorption of reactants or products on the electrode, obviously the equivalent circuit given in Fig. 55 is not adequate. However, one may modify this described method for other equivalent circuits and use it for many kinds of electrode reactions.

D. Steady state measurements

The techniques discussed under this heading will be voltammetry in stirred solutions and polarography. It has become common to refer to the study of current-potential curves at the d.m.e. as polarography and to that at solid eloctrodes as voltammetry.

The essence of steady state measurements is that the function measured (current) is only a function of potential and not of time, even though there is no equilibrium at the system under study. Consider the reaction scheme

$$\text{Ox (in bulk)} = \text{Ox (at electrode surface)} \qquad \qquad \text{I}$$

$$\text{Ox} + \text{ne} \quad = \text{Red} \qquad \qquad \text{II}$$

$$\text{Red} + \text{A} \quad = \text{Product} \qquad \qquad \text{III}$$

where A is a constituent of the reaction solution. If step II is the slowest and step III rapid and irreversible, then the rate of the whole reaction scheme will be determined by the rate constant for step II. Step III will proceed at a constant rate which depends on the supply of reactant Red. There is no contribution from the rate constants of step III to the overall rate. This is a steady state case.

Consider on the other hand a similar reaction scheme where all steps are rapid and reversible. The mass transport step will be in equilibrium and the rate of electron transfer will be very rapid, so that a quasi-equilibrium (as was discussed previously) is maintained and step III, the so called "chemical step", will also be in equilibrium. The whole process will be a quasi-equilibrium one. This quasi-equilibrium process is the polarographically reversible process. Presently we shall describe two ways by which reversibility can be determined and continue to describe some of the experimental aspects of polarography and voltammetry in stirred solutions, devoting a section to the various uses of the rotating disk electrode.

Figures 9 and 13 show current potential curves for reversible and steady state processes. The curves were calculated from the Nernst and Butler–Volmer equations, respectively. If an experimental curve is adequately described by the Nernst equation, this is a reversible reaction. However, it is not convenient to check the current potential curve itself for reversibility, it is much easier to replot it according to equation 3.43. The potential is plotted on the abscissa and $\log (i_1 - i)/i$ on the ordinate. If a straight line is obtained, then the process is reversible. Figure 57 shows a current potential curve for the reduction of Cd^{2+} at the d.m.e. together with the linear log plot. One sees that this reduction is reversible. Figure 58 shows the same data for the polarographic reduction of Ni^{2+}. This reduction is obviously not reversible.

Another convenient way of determining the reversibility of reactions is to study the reduction curves of the oxidized form and the oxidation curves for the reduced form. The half-wave potentials, $E_{\frac{1}{2}}$, of the two curves are identical if the reaction is reversible and the technique is a true steady-state

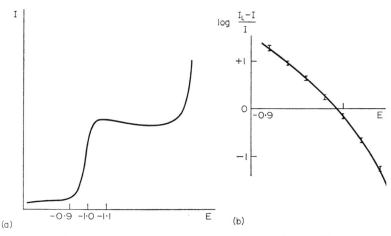

(a)   (b)

FIG. 58. Ni^{2+} reduction at the d.m.e.; (a) Current potential curve; (b) Log plot.

one. In techniques which are not true steady-state, such as voltammetry in quiescent solutions, the peak potential and half-peak-potential (see Chapter 3, Section C) shift somewhat from $E_{\frac{1}{2}}$. This shift was calculated to be

$$E_p = E_{\frac{1}{2}} - \frac{0 \cdot 029}{n} b \qquad\qquad 5.18$$

$$E_{p/2} = E_{\frac{1}{2}} + \frac{0 \cdot 028}{n} b \qquad\qquad 5.19$$

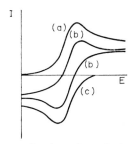

FIG. 59. Current potential curves for the quinone-hydroquinone couple at the graphite electrode; (a) 10^{-3} M quinone, (b) 5×10^{-4} M quinone $+ 5 \times 10^{-4}$ M hydroquinone, (c) 10^{-3} M hydroquinone.

10

(b being the rate of polarization) for a reversible process. This is shown by current potential curves for the system quinone-hydroquinone, taken with the graphite electrode, in Fig. 59. This topic of reversibility will be taken up again when we shall deal with transient measurements, particularly with cyclic voltammetry.

Let us now consider voltammetry in stirred solutions and polarography as opposed to voltammetry in quiescent solutions and understand why the first two methods are steady-state but the third is not. It has been established before that the thickness of the diffusion layer for stationary electrodes in quiescent solutions increases with time. Thus, any method employing these, such as voltammetry, measures current as a function of potential and of time (or polarization rate, which is the same thing) and cannot be regarded as a truly steady-state method. When the solution is stirred, however, the thickness of the diffusion layer reaches a constant value and the measured current is no longer a function of time. Polarography is a steady-state technique because each drop duplicates the exact situation at the previous one—the falling drop stirs the solution enough so that the new drop does not "see" any concentration gradients which have influenced the previous drop.

Polarography occupies a very special place in the history and application of electrode processes. It is with polarography that the interest in the area of electrodics became widespread. Polarography is the most useful of all electroanalytical techniques for the determination of concentrations of metal ions in solution in the millimolar range. The polarographic reduction of oxygen, shown in Fig. 51, will serve to illustrate the scope and some of the limitations and problems of polarography. One sees from Fig. 45(a) that at potentials of ca. zero volts against the s.c.e. the mercury electrode is oxidized, preventing the study of electrode processes at more positive potentials. The increase in current at very negative potentials is due almost always to the reduction of the supporting electrolyte, only when the supporting electrolyte is a tetra alkyl ammonium salt does this rise correspond to the reduction of water. This reduction of water takes place more than 2·5 volts negative to the s.c.e. Reactions which take place at potentials more negative than that cannot, of course, be studied, but the number of these reactions is very small. Very often one does not need potentials higher than 1·5 or so volts negative to the s.c.e.

The solutions which are suitable for polarography are typically made up of a conducting medium (such as 0·1 M solution of KCl in water) and the electroactive species in millimolar concentration. When the concentration drops below 10^{-4} M, the current is often too small to measure and when the concentration is much higher than 2×10^{-2} M the concentration of the supporting electrolyte may not be large enough to eliminate electric migration as a form of mass transport of the electroactive species. Another problem associated with very concentrated solutions is that of maxima which will be dealt with shortly.

One sees that the polarographic curve is not smooth but consists of up-surges and drops of the current. This form is a result of the use of the d.m.e. Each drop grows and the current increases as the area increases; when the drop falls and the effective area of the electrode becomes the area of the capillary, the current drops and then rises again as the other drop grows.

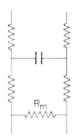

FIG. 60. A damping circuit; R_m is the measuring resistance.

Sharp variations in the current are rather inconvenient and, therefore, a damping device is introduced into every polarographic circuit. The design of a typical damping device is shown in Fig. 60. In effect all it does is to increase the time of response of the recorder by letting the sharply variable current pass through a capacitor before it reaches the recorder circuit.

Figure 51 shows a very sharp increase of current at the first wave of oxygen. As the potential is increased this current falls to its normal value, producing a maximum. Two kinds of maxima are distinguished, the one in Fig. 51 being of the first kind. Maxima of the first kind are associated with the reduction of inorganic materials mainly and are of the characteristic spike shape. Maxima of the second kind appear usually with reactions of organic compounds, usually in the vicinity of the point of zero charge and are broad and round in shape. Both kinds of maxima can be suppressed or eliminated by adding small amounts of specifically adsorbed substances, called "maximum suppressors", the most common being gelatin. Concentration of about 0·01 % is adequate for the complete elimination of the maximum. Figure 61 shows the influence of several concentrations of gelatin on the reduction wave of copper in nitrate medium. High concentrations of gelatin are undesirable since they may suppress the whole wave or distort its shape. Maxima, especially of the second kind, are more pronounced in solutions of high concentration; if the concentration of the supporting electrolyte is decreased, the maximum may be eliminated without the use of maximum· suppressors.

Maxima have been shown to be associated with streaming at the mercury drop surface, perhaps because the potential at some part of the surface differs from that at the other part. However, the reason for this streaming and its cessation in the presence of maximum suppressors has not been well explained.

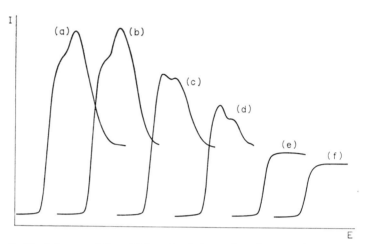

Fig. 61. The influence of the addition of gelatin on the polarographic wave of copper; (a) No gelatin; (b) 1 drop; (c) 5 drops; (d) 10 drops; (e) 20 drops; (f) 40 drops.

Mercury has the property of dissolving many metals, forming amalgams. If the mercury is not properly purified before being used for the d.m.e. the base curve, or residual current curve, will show a small oxidation current, which means that an oxidation process is taking place. Obviously, reduction waves superimposed on such residual current curves will be too low.

The most common form of stirring in voltammetry with stirred solutions is by the use of the rotating disk electrode. Figure 62 shows a design of this electrode. The active surface is embedded in a large disk of Teflon shaped in the form shown. This particular form has been shown to give the least interaction between that part of the solution which is under the surface of the

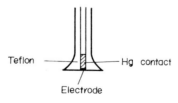

Fig. 62. A rotating disk electrode.

disk and that which is above the surface of the disk, making the disk area more "infinite". Electrical connection is made with mercury put into a hole in the Teflon sheath or by direct soldering of the electric cable to the electrode. In the latter case some form of transfer of electricity from the rotating cable to the fixed instrument must be designed.

The great usefulness of the rotating disk electrode is that the mass transport controlled current is proportional to the square root of the rotation speed. If the rate of the electrode process is controlled by both the rate of electron transfer and of mass transport, it can be shown that the current flowing through the cell is related to the rotation speed by the following equation

$$\frac{1}{i} = \frac{1}{i_{max}} + \frac{1}{\omega^{\frac{1}{2}}} f(D_R, D_{Ox}, \beta, \eta) \qquad 5.20$$

where ω is the rate of rotation and

$$i_{max} = i_0 \left(\exp \beta \frac{nF}{RT} \eta - \exp - (1 - \beta) \frac{nF}{RT} \eta \right)$$

A graph of $1/i$ against $1/\omega^{\frac{1}{2}}$ gives a straight line, the intercept of which gives information on the kinetics of the electrode reaction studied. However, when i_0 is very large, the intercept becomes very small, decreasing the accuracy of the method. Obviously, if the reaction can be studied at very high rotation speeds, a better value will be obtained for i_{max}. Since very high speeds of rotation can cause turbulence and cavitation, great care should be taken in the design of the rotating parts and the placement of the electrode in the solution.

A special form of a rotating electrode is the ring-disk electrode, a schematic diagram of which is shown in Fig. 63. The electrode is constructed from a centre disk which is conducting and is connected to its own voltage source and recorder. The ring, which is also conducting, is separated from the disk

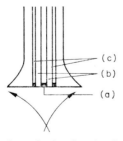

Fig. 63. A rotating ring disk electrode showing the direction of flow of liquid; (a) The disk; (b) The insulating gap; (c) The ring.

by an insulating gap and is connected to a different voltage source and recorder, making the ring-disk assembly like two combined measuring instruments. The direction of flow of the solution close to the surface of the disk is radial, any intermediate or reaction product formed on the disk will be carried towards the ring. Consider now the two reaction schemes

$$Red = Ox + ne \qquad \qquad \text{IV}$$

and

$$Red = Ox + ne \qquad \qquad \text{V}$$

$$Ox \overset{k}{=} A + B \qquad \qquad \text{VI}$$

Reaction IV takes place at the disk and the ring is held in a potential such that $[Ox]_r = 0$, i.e. the product of reaction IV is reduced at the ring. Under such arrangement the ring current I_r will obviously differ from the disk current because some of Ox never reaches the ring surface. The collection efficiency N is defined

$$N = \frac{|I_r|}{|I_d|} \qquad \qquad 5.21$$

and depends in a complex, but known, way on the radius of the disk and the thicknesses of the gap and the ring. The experimental determination of N is fairly straightforward and a number of reversible redox reactions can be used.

The great attraction and power of the rotating ring-disk system is, however, in the discovery of intermediates, such as in reaction schemes V and VI. If the potential at the disk is maintained at a value such that reaction V proceeds from left to right and the ring potential is maintained so that the reverse reaction will occur (i.e. reaction V from right to left), the ring current will obviously be less than $N|I_d|$ since part of Ox reacts via VI. This deficiency in the ring current is related to the rate constants of reaction VI and can be used for the evaluation of these rate constants. If, on the other hand, A or B are electroactive, the rise in their concentration can be observed via the ring current, thus, again enabling the evaluation of the rate constants of reaction VI. The number of possibilities of detection of intermediates and evaluation of rate constants is limited only by the potential range of the electrode material and by the skill of the operator.

The most used material for solid electrodes has been platinum, mainly because of its chemical inactivity. Platinum, however, has proved to give a rather limited range of potential in which it is truly polarizable. Figure 64 shows curves taken with a platinum electrode in $1M$ H_2SO_4. The anodic part of the curve (positive current) shows an increase in current at the potential of $0 \cdot 8$ V. This increase is due to oxidation of the surface of the platinum electrode. The reduction of the surface oxidation product is evident from the increase of cathodic current; at about $0 \cdot 3$ V another cathodic process takes place, the reduction of hydrogen ions. Hydrogen gas is adsorbed on the surface of the platinum and is reoxidised when the potential is reversed again. Thus, when working with platinum one must

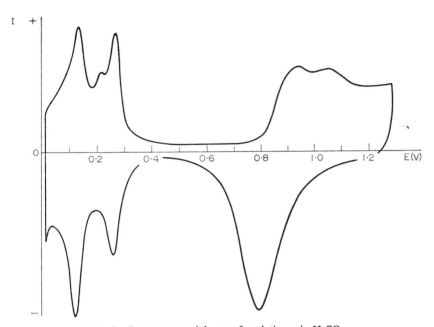

FIG. 64. Current potential curve for platinum in H_2SO_4.

be aware of surface oxidation or adsorption of hydrogen though not all supporting electrolytes will give the same curves as sulphuric acid. In some electrolytes the potential range available is larger, in other electrolytes, notably those containing halide, CN^- or CNS^- ions, the range is smaller due to the formation of platinate complexes.

The cleaning of the electrode surface presents a very difficult problem with the platinum electrodes. Some workers recommend cleaning with dilute sulphuric acid solution, others recommend treatment with chromic acid (which, of course, oxidizes the electrode surface). Still other workers recommend "pretreatment" at a constant potential for a certain length of time or at a series of potentials. The best way to be followed seems to be the one best suited to the problem at hand in that logical and reproducible results are obtained. A method of pretreatment, however, must be devised before one is able to produce data to any degree of reliability.

Another popular material for the construction of solid electrodes is graphite. There are several kinds of graphite which differ significantly from each other as to the mechanical form, porosity and degree of purity. Some of the most useful forms for making electrodes are the spectroscopic graphite rods and pure graphite powder. The former are usually treated with wax under vacuum in order to plug the pores and reduce the roughness of the graphite surface. The powder is mixed with a suitable organic binder, such as Nujol, to a paste and is put into a specially designed holder, such as that shown in Fig. 65. The two other forms of graphite which have become useful

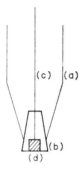

FIG. 65. A holder for graphite paste electrodes; (a) A standard taper glass joint; (b) A Teflon plug; (c) A platinum or copper wire; (d) The carbon paste.

recently are the pyrolitic graphite and glassy carbon. Pyrolytic graphite is manufactured by pyrolysis of carbonaceous gases at reduced pressure and is a form of highly organized material, some specimens of this graphite are virtually single crystals. This material exhibits a pronounced anisotropy in conductance of both heat and electricity and its resistance to mechanical manipulation is also quite different along the two directions since it flakes very easily in one direction only, making its machining rather difficult. Its main advantage as electrode material lies in its high purity and defined structure. The glassy carbon is a form of graphite which resembles glass in its hardness and brittleness. It is isotropic and is claimed to be very suitable as electrode material.

Surface preparation of graphite is less defined than that for platinum. Here, in addition to the various cleaning methods described previously, mechanical methods of surface renewal can be used, i.e. the wax impregnated electrode can be cut on a lathe and a fresh surface prepared for every experiment. The carbon paste electrode can be totally renewed for every experiment.

After having described one equilibrium method for the measurement of kinetics of rather fast electrode processes and two steady state methods for measurement of slow electrode processes, let us turn now to consider the so called "transient" or, sometimes, "relaxation" techniques, where time is a very important factor in the equations.

E. Transient measurements

All transient techniques can be classified into two broad groups. The first group includes those where a known time function of the electrical variable is applied to the electrode and the dependent variables are then measured as a function of time. The methods of this group can further be subdivided according to the independent electrical variable which can be either potential, current or charge and according to the function of this variable with time: pulse or ramp. The methods of this first group usually yields quantitative information on the kinetics of the reaction studied, they measure i_0 and β.

The second group of techniques can be broadly called "cyclic voltammetry". Here a potential ramp is applied to the electrode, which is either solid or a mercury drop (not the d.m.e.) and the resulting current is measured. The current-potential curve obtained is usually of the form given in Fig. 11. Very often one reverses the potential sweep and records the current observed for the reverse process. Typical current potential curves for cyclic voltammetry are given in Fig. 66.

Let us start our discussion of transient measurement with cyclic voltammetry. The cell used is identical with that used for regular voltammetry or polarography, the set-up of the instruments is also identical except for the voltage source. The voltage source used in this technique provides a voltage ramp of known slope, i.e. the rate of increase of potential with time (known

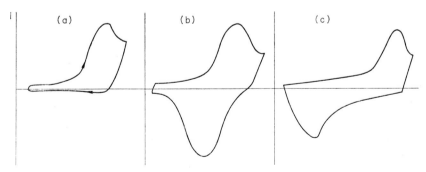

FIG. 66. Typical cyclic voltammograms for: (a) An irreversible process; (b) A reversible process with a fast electron transfer step; (c) A reversible process with a slow electron transfer step.

as the "polarization rate") is known and adjustable. It also provides a means of reversing the direction of polarization at any fixed point on the potential scale. Sometimes motor driven potentiometers are used to provide the potential ramp, sometimes operational amplifiers are used.

Consider the following reaction scheme

$$Red = Ox + ne \qquad\qquad VII$$

$$Ox + solvent = product \qquad\qquad VIII$$

and Fig. 66. Assume that the potential is being scanned from a value where reaction VII does not occur to that where the reaction is diffusion controlled and then back to the starting value. As the oxidation potential of Red is reached, Ox molecules are being formed near the electrode surface. These will react at a certain rate with solvent molecules to produce the reaction products. While Ox is being reacted with the solvent, the potential may reach its final value and start in the reverse direction again, reaching the reduction potential of Ox. If the rate of reaction VIII is much faster than the rate of polarization, no reduction wave will be observed in the reverse scan [Fig. 66(a)]. If the rate of reaction VIII is slow compared to the rate of polarization, the reduction wave of Ox will be observed. Thus it may actually be seen that a curve which looked very much like that of Fig. 66(a) for a slow polarization rate, will transform to curve (b) at a high polarization rate.

When an electrode process VII takes place, but the reaction is not polarographically reversible, one may observe a cyclic voltammogram as in Fig. 66(c). The separation of the peaks depends on the polarization rate: it is larger for higher polarization rates. For a given polarization rate, however, the separation of the peaks indicates the value of k_s—the larger k_s, the smaller the separation.

The shape of the current potential curves of successive cycles is not expected to be identical because the concentrations of the reactants and products near the electrode at the end of a cycle are not identical to that at the beginning of the experiment. Moreover, the shape of successive cycles sometimes reveals electroactive reaction intermediates which cannot otherwise be detected. The literature on cyclic voltammetry is extensive and varied, organic and inorganic reactions have been studied in a variety of solvents. Often cyclic voltammetry is used initially to explore qualitatively the reaction under study before undertaking quantitative measurements of its kinetics.

In the following we shall deal with several of the transient methods which are commonly used for the determination of the rate constants of electrode reactions. A few other known methods will not be described here in detail either because they are obsolete, involve rather uncommon electronic equipment or are very difficult to analyse mathematically, adding nothing but complexity to our experimental possibilities.

The simplest transient method for the study of rapid electrode processes is the voltage step method. It consists of applying a sudden voltage step to an electrochemical cell initially at equilibrium. This voltage step is applied across the cathode and anode and the current passing through the cell is

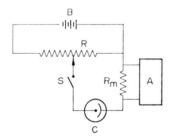

FIG. 67. A schematic diagram of the circuit for the potential step method; A: oscilloscope; B: battery; C: cell; R: potentiometer; R_m: measuring resistance; S: switch.

recorded as a function of time. A schematic diagram of the circuit to be used is given in Fig. 67. The potential of the working electrode, however, is not the same as the voltage step V because it includes the IR drop in the cell

$$V = \eta + IR_s \qquad 5.22$$

R_s is the resistance of the cell solution. The shapes of the working electrode potential and the current against time curves are shown in Fig. 68.

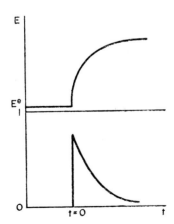

FIG. 68. Potential and current-time curves of the voltage step method.

Let us now consider a simple, oxidation reaction, such as in VII. When the reaction is very fast we cannot neglect the contribution of diffusion to the overall overpotential, even when the latter is only a few millivolts. The Butler–Volmer equation for the case where the concentrations of the reactants at the electrode surface differ from those in the bulk is

$$i = i_0 \left\{ \frac{[R]}{[R]_0} \exp \left[\beta \frac{nF}{RT} \eta \right] - \frac{[Ox]}{[Ox]_0} \exp - \left[(1 - \beta) \frac{nF}{RT} \eta \right] \right\} \quad 5.23$$

Where, as before the subscript 0 denotes the concentration in the bulk of solution and no subscript denotes the concentrations at the electrode surface. If the overpotential is very small $(2-5\,\text{mV})$, one can linearize equation 5.23 to obtain

$$i = i_0 \left(\frac{[R]}{[R]_0} - \frac{[Ox]}{[Ox]_0} - \frac{nF\eta}{RT} \right) \quad 5.24$$

The linearization of equation 5.23 is not straightforward, but the exact way in which this is done will not be shown here. Equation 5.24 contains two unknowns, i and E. In order to find them one must solve Fick's law with the boundary conditions 3.45 to 3.47. Equation 5.24 is the fourth boundary condition. The result is

$$\left. \begin{array}{l} i = \dfrac{i_0 \, nFV}{i_0 \, R_s nF + RT} \exp Q^2 \, t \operatorname{erfc} Qt^{\frac{1}{2}} \\[3mm] Q = \dfrac{RT}{nF} \dfrac{i_0}{i_0 \, R_s nF + RT} \left(\dfrac{1}{[Ox]_0 \, D_{Ox}^{\frac{1}{2}}} + \dfrac{1}{[R]_0 \, D_R^{\frac{1}{2}}} \right) \end{array} \right\} \quad 5.25$$

The current density i depends on time via the function $\exp \xi^2 \operatorname{erfc} \xi$. erfc is the complementary error function

$$\operatorname{erfc} \xi = 1 - \frac{2}{\pi^{\frac{1}{2}}} \int_0^{\xi} \exp - z^2 \, dz \quad 5.26$$

which varies from one, where $\xi = 0$, to zero when $\xi \to \infty$. The function $\exp \xi^2 \operatorname{erfc} \xi$ also varies from one to zero and is shown in Fig. 69. This function can be linearized for values $Qt^{\frac{1}{2}} \ll 1$

$$i = \frac{i_0 \, nFV}{i_0 \, R_s nF + RT} \left(1 - \frac{2Qt^{\frac{1}{2}}}{\pi^{\frac{1}{2}}} \right) \quad 5.27$$

If one plots the observed current against $t^{\frac{1}{2}}$ at sufficiently small times, one gets a straight line and can calculate i_0 from the intercept J.

$$J = \frac{i_0\, nFV}{i_0\, R_s\, nF + RT}$$

$$i_0 = \frac{JRT}{nFV - JR_s\, nF} \qquad 5.28$$

erfc ξ expξ^2

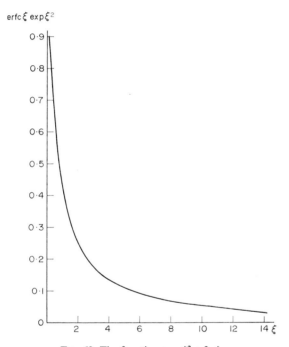

FIG. 69. The function $\exp \xi^2$ erfc ξ.

R_s must be measured independently. However, at very short times the current passed through the cell is a capacity current, i_C and not the Faradaic current.

$$i_C = \frac{V}{R_s}\, \exp - \frac{t}{R_s\, C} \qquad 5.29$$

Thus it is necessary to choose a time t such that the capacity current is small and the faradaic current is still given by equation 5.27. The correct times to be used are estimated by equations 5.29, 5.27 and 5.25. Values for the function $\exp \xi^2$ erfc ξ are given in the literature.†

† This function is also encountered in heat conduction in solids and can be found in some collections of mathematical tables.

t usually must be of the order of 10^{-4} sec. Variations over more than one order of magnitude are not usually permissible. Let us now discuss the scope of the voltage-step method, i.e. what is the largest value of i_0 which it will measure. We assume that if the slope of the η vs. $t^{\frac{1}{2}}$ curve is larger than unity, the error involved in measuring the intercept and, hence, i_0 is too large. Operationally it is difficult to measure the intercept on curves which are much steeper than this. The slope is given by equation 5.27

$$\text{slope} = 1 = \frac{i_0\, nF}{i_0\, R_s\, nF + RT}\, \frac{V}{}\, \frac{2Q}{\pi^{\frac{1}{2}}} \qquad 5.30$$

Q, in turn, is given by equation 5.25. We assume $[\text{Ox}]_0 = [R]_0 = 1$ mole m^{-3} (10^{-6} mole cm^{-3}), $D_R = D_0 = 10^{-9}$ m^2 sec^{-1} (10^{-5} cm^2 sec^{-1}), $R_s = 50\Omega$, $n = 1$ and $T = 300°$K. The result of the calculation is that

$$i_0 \leqslant 10^2 \text{ A m}^{-2} \ (10^{-2} \text{ A cm}^{-2})$$

in order for the voltage step method to apply. According to equation 3.68

$$i_0 = nFk_s[\text{Ox}]_0^{\beta}\,[R]_0^{1-\beta}$$

and the same concentrations

$$k_s \leqslant 10^{-3} \text{ m s}^{-1} \ (10^{-1} \text{ cm s}^{-1})$$

If a potentiostat is available the potential step method may be used where the potential of the working electrode, rather than the cell voltage, is changed suddenly and kept constant. A schematic diagram of the circuit and the current and potential against time curves are given in Figs 70 and 71. The results of solving Fick's diffusion equation and the appropriate boundary conditions is the same as the voltage step method when $R_s = 0$. The same plot of η vs. $t^{\frac{1}{2}}$ is made and the intercept is

$$J = \frac{nF}{RT}\, i_0\, E \qquad 5.31$$

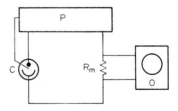

Fig. 70. A schematic diagram for the potential step method; P: potentiostat; C: three electrode cell; R_m: measuring resistor; O: oscilloscope.

where E is the height of the potential step.

$$i_0 = \frac{RT}{nF}\frac{J}{E} \qquad\qquad 5.32$$

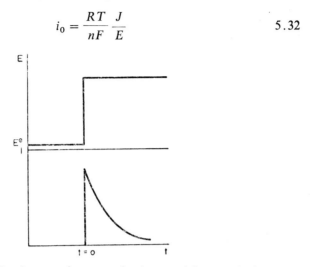

FIG. 71. Potential and current-time curves for the potential step method.

Equation 5.29 predicts that for $R_s = 0$, as is the case with the potential step method, one does not have to worry about the capacity current. Therefore the time of measurement can be as short as practicable, contrasting with the case of the voltage step method. Using the same criterion of the maximum i_0 that can be conveniently measured as for the voltage step method, we get for the slope

$$\left.\begin{array}{c} \text{slope} = 1 = \dfrac{nF}{RT} i_0 E \dfrac{2Q}{\pi^{\frac{1}{2}}} \\[2em] Q = \dfrac{i_0}{nF}\left(\dfrac{1}{[Ox]_0 D_{Ox}^{\frac{1}{2}}} + \dfrac{1}{[R]_0 D_R^{\frac{1}{2}}} \right) \end{array}\right\} \qquad 5.33$$

Using the same values as before we get

$$i_0 \leqslant 8 \times 10^2 \, \text{A m}^{-2} (8 \times 10^{-2} \, \text{A cm}^{-2})$$

which means that the potentiostatic method is capable of handling reactions which are faster by almost an order of magnitude than the voltage step method.

Let us now discuss those methods where the current is the independent variable. Two experimental methods will be discussed, one where one current step is applied to the cell and the other where two current steps are applied.

The basic circuit for the simple current step technique is given in Fig. 72, and the resulting current and potential-time curves are given in Fig. 73. The battery B should give a fairly high voltage and the resistor R should be much larger than the total cell resistance R_s, so that the current $I = V/(R+R_s)$

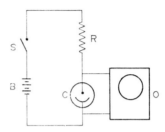

FIG. 72. A schematic circuit for the simple galvanostatic technique; B: battery; S: switch; R: large resistance; C: cell; O: oscilloscope.

would not be influenced by R_s. In order to minimize the IR drop, the cell can be incorporated into a bridge circuit as shown in Fig. 74. Here $R_1 = R_2$; therefore, when $\eta = 0$ and the bridge is balanced, $R_3 = R_s$. Thus, when the current step is applied, the oscilloscope will register η directly without an IR component.

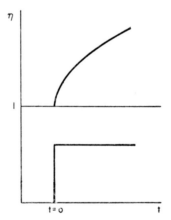

FIG. 73. The dependence of current and of cell potential on time for the galvanostatic method.

The constant current applied from the generator is the sum of the capacity and faradaic currents

$$i = C\frac{d\eta}{dt} + nFD\left(\frac{\partial C}{\partial x}\right)_{x=0} = \text{constant} \qquad 5.34$$

where c is the concentration of the reactant. This equation is now a boundary condition for Fick's law. Another boundary condition is equation 5.23. An explicit solution of this problem (equations 3.22, 3.45, 3.46, 3.47, 5.23 and 5.34) is not possible and therefore equation 5.24 replaces 5.23 as the

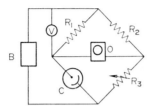

FIG. 74. A schematic diagram for a bridge set-up for the current step method; this set-up eliminates IR_s drop from the measured overpotential as read on the oscilloscope; B: a pulse generator; C: the cell; V: voltmeter for monitoring the current flow; R_1 and R_2: equal fixed resistors; R_3: variable resistor; O: oscilloscope.

boundary condition. This means that η must be smaller than 5 mv. The overpotential-time curve is given, under this condition, by

$$\eta = \frac{RT}{nF}\, i\, \left[\left(\frac{1}{[Ox]_0\, D_{Ox}^{\frac{1}{2}}} + \frac{1}{[R]_0\, D_R^{\frac{1}{2}}}\right)\frac{2t^{\frac{1}{2}}}{nF\pi^{\frac{1}{2}}} - \left(\frac{1}{[Ox]_0\, D_{Ox}^{\frac{1}{2}}} + \frac{1}{[R]_0\, D_R^{\frac{1}{2}}}\right)^2 \frac{RT}{n^3\, F^3}\, C + \frac{1}{i_0}\right] \quad 5.35$$

η is plotted against $t^{\frac{1}{2}}$ and i_0 is calculated from the intercept J for $t = 0$. C, the double-layer capacity, must be measured separately.

$$J = \frac{RT}{n^2\, F^2}\left(\frac{1}{[Ox]_0\, D_{Ox}^{\frac{1}{2}}} + \frac{1}{[R]_0\, D_R^{\frac{1}{2}}}\right)iC + \frac{RT}{nF}\frac{i}{i_0} \quad 5.36$$

The intercept is composed from a capacitive term and an electron-transfer term. If the evaluated i_0 is to be meaningful, the second term should not be much smaller than the first. Suppose that the second term is one-fifth of the first and calculate the resulting i_0 for the following conditions:

$$[Ox]_0 = [R]_0 = 1\ \text{mole m}^{-3};$$

$$D_{Ox} = D_R = 10^{-9}\ \text{m}^2\ \text{sec}^{-1};$$

$$i = 0{\cdot}1\ \text{A m}^{-2},\ C = 0{\cdot}1\ \text{F m}^{-2},\ n = 1$$

The result is:

$$i_0 \sim 5 \times 10^3\ \text{A m}^{-2}\quad (0{\cdot}5\ \text{A cm}^{-2})$$

Thus, the simple galvanostatic technique is useful for studying reactions an order of magnitude faster than those suitable for the potentiostatic technique.

As stated before, the intercept of the η vs. $t^{\frac{1}{2}}$ curve includes a capacitive term. If it were possible to eliminate this term, even faster reactions could be studied. The capacity current can be virtually eliminated by charging the double layer to the correct overpotential $\left(\eta = (RT/nF)(i/i_0)\right)$ using a short pulse of current, I_1, prior to the "measuring current" I. The shape of this

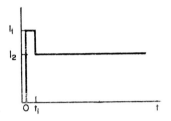

FIG. 75. The dependence of current on time of the double pulse galvanostatic method.

double pulse is shown in Fig. 75. Figure 76 shows the overpotential-time curve for the case where (a) I_1 is too small, (b) I_1 is correct and, (c) I_1 is too large. The time t, must be much smaller than the time of measurement. The correct ratio I_1/I can be found by trial and error.

When studying electrode reactions where i_0 is of the order of 10^4 amperes m^{-2}, one must consider the effect of diffusion even at t_1 (which may be 1-2 μs.). It has been shown that the horizontal part of the η vs. t curve, η_0, is given by

$$\eta_0 \simeq \frac{RTi}{nF} \left[\frac{1}{i_0} + \frac{4}{3\pi^{\frac{1}{2}} nF} \left(\frac{1}{[Ox]_0 D_{Ox}^{\frac{1}{2}}} + \frac{1}{[R]_0 D_R^{\frac{1}{2}}} \right) t_1^{\frac{1}{2}} \right] \qquad 5.37$$

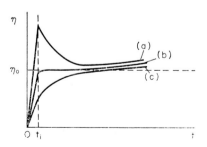

FIG. 76. The dependence of the overpotential on time for various values of I_1; (a) I_1 is too small, (b) I_1 is right, (c) I_1 is too large.

t_1 can be varied by varying I_1 and the duration of the pulse (both of them are independently variable on commercial generators) and an extrapolation of η_0 to $t_1 = 0$ must be done in order to calculate i_0. The reactions studied by this method may be an order of magnitude faster than those studied by the single pulse galvanostatic method.

Another transient method which may be used to study fast electrode processes is the coulostatic method. Here a known amount of electricity is "injected" into the electrode in a very short time. The electrode, initially at

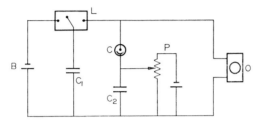

FIG. 77. A schematic circuit for the coulostatic method: B: battery; L: relay switch; C_1, C_2: capacitors; C: cell; O: oscilloscope; P: potentiometer.

the equilibrium potential, acquires an overpotential which then decreases with time as the electrode reaction proceeds. A schematic diagram of the appropriate circuit is shown in Fig. 77. The capacitor C_1 (0·3 μF) is charged to a known extent by the battery B

$$q = C_1 V_B \qquad\qquad 5.38$$

q is the charge on the capacitor, C_1 its capacity and V_B is the battery voltage. After C_1 had been charged, the switch is thrown to disconnect the battery and connect the cell. The potentiometer P and capacitor C_2 are used to neutralize the equilibrium potential so that the oscilloscope would register zero before the start of experiment.

When the switch is connected to the cell, η rises during the first microsecond until it reaches a maximum η_{max}. The time when $\eta = \eta_{max}$ is taken

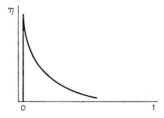

FIG. 78. The dependence of overpotential on time for the coulostatic method.

as zero. A typical curve is shown in Fig. 78. Since the charge q of the capacitor C_1 is transferred to the electrode, the capacity of the double layer is easily found

$$C = \frac{C_1 V_B}{\eta_{max}} \qquad 5.39$$

In order to find the exact form of $\eta - t$ curves, Fick's diffusion equation must be solved together with the appropriate boundary conditions. In order to do that let us denote as q_0 the charge of the electrode at equilibrium, q_{max} the charge of the electrode at η_{max}, i.e.

$$q_{max} = q_0 + q = q_0 + C_1 V_B = q_0 + \eta_{minax} \qquad 5.40$$

but q_{max} approaches q_0 with time because of the electrochemical reaction

$$q = q_{max} + \int_0^t I_f \, dt \qquad 5.41$$

$$\eta = \eta_{max} + \frac{1}{C} \int_0^t I_f \, dt \qquad 5.42$$

η is substituted from equation 5.24 and I_f from equation 3.47 where $I_f = i_f A$. The resulting equation is the necessary boundary condition. The full solution is complicated and never used. However, for the case where

$$\frac{4nF}{RTCi_0} \gg \frac{1}{n^2 F^2} \left(\frac{1}{[Ox]_0 D_{Ox}^{\frac{1}{2}}} + \frac{1}{[R]_0 D_R^{\frac{1}{2}}} \right)^2 \qquad 5.43$$

the following relation is obtained:

$$\eta = \eta_{max} \exp - \frac{i_0 nF}{CRT} t \qquad 5.44$$

Thus, a plot of $\log \eta$ against t should yield a straight line. Extrapolation to $t = 0$ yields η_{max} and, consequently C. i_0 is obtained from the slope of the curve. The use of the method is limited by equation 5.43. If the left-hand side of expression 5.43 is fifty times the right-hand side, we obtain (for the same values as before)

$$i_0 = 80 \, A \, m^{-2} \quad (8 \times 10^{-3} \, A \, cm^{-2})$$

as the value of exchange current density which can be measured with this technique. The disadvantage of not being able to determine high i_0 values is sometimes compensated by the simplicity of the apparatus required.

All methods for the determination of electrode kinetics discussed here cover a very wide range of exchange current density values. The approximate value for i_0 can be estimated from cyclic voltammetry and the appropriate technique can then be chosen for further study. However, the choice of technique may not be determined by the value of i_0 alone, the availability of instruments may influence the decision as well as other factors, e.g. extreme conditions of temperature and pressure or electrode configuration.

F. Reflection spectroscopy and ellipsometry

All techniques discussed so far can conveniently be called "electronic" in that they use the electrical characteristics of the electrode reaction in order to learn about its kinetics and mechanism. Many electrode reactions are strongly influenced by adsorbed films; information on adsorbed films is available from capacity measurements, but this information is indirect and, especially with solid electrodes, not very accurate. Clearly, more direct information on adsorbed films is desirable.

Another area where the electronic methods are not very good is in the identification of reaction intermediates. The only tools for detection of intermediates are the ring-disk electrode and cyclic voltammetry; clearly another method for the detection and determination of intermediates is needed.

Various types of spectroscopy have been used very successfully for the determination of structure of molecules and intermediates in solution. However, the application of these techniques to electrode processes is limited by the fact that the latter take place at the surface and direct spectroscopy is clearly not possible since the electrode is opaque. There is one exception in that e.s.r. spectroscopy has been used for some time in identifying radicals as intermediates in electrode reactions. In order to do this an electrochemical cell must be constructed in the e.s.r. cavity. This has been done and some very interesting data on the mechanism of several organic electrode reactions resulted. However, the application of e.s.r. spectroscopy to the study of electrode processes is limited to free radical intermediates the concentration of which is fairly high, i.e. the life time is fairly long.

In the last few years there has been an increased interest in the application of reflectance spectroscopy to the study of electrode processes. Initially this was hampered by the fact that the change in absorbing species is confined to a very thin layer near the electrode, making the length of the beam path in the absorbing medium very short. A short path results in very small absorbance, except in cases where the absorption coefficient is extremely high. Indeed, early studies of electrode reactions with reflectance spectroscopy dealt with reductions and oxidations of diestuffs. This problem is now over-come by the use of very sensitive detectors.

The area of spectroscopic studies of the electrode-solution interphase can be divided into four parts according to the four techniques used. In the following each technique will be briefly described and its merits and limit-ations will be discussed.

Attenuated Total Reflectance (A.T.R.) will be the first technique to be described. If we take a plate of material of high refractive index n_2 so that the refractive index of the plate is higher than that of air (n_1) or than that of

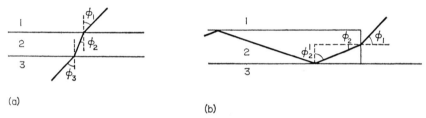

(a)

(b)

FIG. 79. The attenuated total reflection plate.

the solution to be studied (n_3), we may shine a beam of light at an angle so that the beam is totally reflected from the surface of the plate, as shown in Fig. 79. Figure 79(a) shows that the angle ϕ_3 is given by

$$\sin \phi_3 = \frac{n_1 \sin \phi_1}{n_3} \qquad 5.45$$

so that if $n_3 > n_1$, i.e. the index of refraction of the solution is greater than that of air, $\sin \phi_3$ can never equal one, i.e. there can never be a total reflection from the plate-solution interface. Therefore, the light beam has to shine from the side of the plate as shown in Fig. 79(b). The critical angle ϕ_{1c}, for total reflection is obtained from Snell's law

$$\sin \phi_{1c} = \frac{(n_2{}^2 - n_3{}^2)^{\frac{1}{2}}}{n_1} \qquad 5.46$$

The intensity of the reflected beam of light is very near the intensity of the incident beam provided that the solution does not absorb in the wavelength used. If the solution does absorb, the intensity of the reflected beam will be smaller. As a range of wavelength is scanned, a spectrum of the solution is obtained which is quite similar to the usual absorption spectrum. This means that the light beam, as it is being reflected from the surface of the plate, penetrates the solution to some extent, giving information on any absorbing species that may be present at or near the surface of the plate. If the plate is coated with a semitransparent layer of metal, the latter can be used as an electrode and spectra of intermediates and products can be thus recorded.

Gold has been used extensively as the metal which was plated on a glass plate. This arrangement obviously is limited to the visible range. The technique was extended to the infrared range by using germanium plates which serve as both electrodes and plates. The scope of this technique seems to be very large, but to date only a few studies have used it.

The second technique is that of transmission spectroscopy through transparent electrodes. The principle of this technique is the same as that of absorbtion spectroscopy whereby if an absorbing substance is present in the diffusion layer near a transparent electrode its spectrum could be recorded in the usual way. The problem with this technique is obviously the need for a transparent electrode. Germanium and doped tin oxide can be used, as can a very thin layer of metal deposited on glass or silica plates. The thin layer can be reduced to a grid which will give conductance to the surface of the plate without making it too opaque.

The third method is that called Specular Reflectance Spectroscopy. A schematic diagram of the spectroscopic part of the experimental set-up is shown in Fig. 80. Here the electrode is made of a reflecting metal, commonly

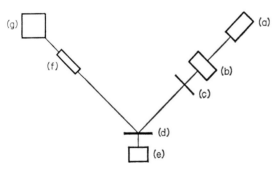

FIG. 80. Specular reflection spectrophotometer; (a) Light source; (b) Monochromator; (c) Polarizer; (d) Reflecting metal; (e) Voltammetric circuit with modulator; (f) Photo cathode or photomultiplier; (g) Phase sensitive detector.

platinum, which is polished to give the best possible reflecting plane. The incident light beam is usually plane polarized to give a higher sensitivity to the measurement and the intensity of the reflected beam is amplified and recorded. The potential of the electrode is modulated, i.e. changed from zero to the value E and back again, at a prescribed frequency. Thus, the absorbing species appears and disappears, making the light intensity change from maximum intensity to maximum absorption for that potential at that frequency. The electric current generated by the photomultiplier acquires an a.c. component, which is filtered and measured by the phase sensitive detector. Species in solution which absorb at the same wavelength as the one present at the electrode surface, do not affect the results since their absorption does not give rise to an a.c. current. Initially it was thought that many reflections should give a better sensitivity to the method, but later it was discovered that dispersion of the reflected beam increases very much on multiple reflections. If a beam of light is reflected from an ideal flat surface, the area of the reflected beam is the same as that of the incident beam. If, however, the reflecting surface is not perfectly flat, the area of the reflected beam is larger than that of the incident one. This effect is multiplied when there are several reflections with the result that the area of the beam reaching the photomultiplier is larger than the area of the slit used in the optical path so that the intensity actually recorded is much less than the true one.

This technique has been used for the study of adsorbed layers on metals and intermediates of electrode reactions. Again, the full potentialities of the technique have not yet been realized.

The last technique to be described here is ellipsometry. This has been used for many years by physicists to study adsorbed films on a large variety of surfaces. Recently it gained some popularity with surface chemists, especially surfaces. Recently it gained some popularity with surface chemists. The experimental set-up is somewhat similar to that used for specular reflection spectroscopy but differs from it in one important aspect which is a result of the parameters measured in ellipsometry. A schematic diagram of the optical part of an ellipsometer is shown in Fig. 81. Plane polarized light of a definite wavelength (this is the crucial difference) shines on the electrode surface and

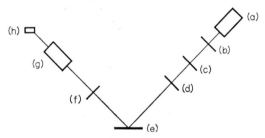

Fig. 81. Schematic presentation of an ellipsometer; (a) Light source; (b) Filter; (c) Polarizer; (d) Quater wave plate (compensator); (e) Electrode; (f) Analyser; (g) Collimating lenses; (h) Detector.

is reflected from it. On reflection it becomes elliptically polarized in a way which is characteristic of the reflecting surface. The elliptically polarized light is converted to plane polarized by a "quater wave plate", but the angle of polarization of the latter beam differs from that of the original beam. This angle is measured by the analyser which is rotated to give minimum light intensity. The polarizer and the analyser are built into divided circles, the change of the angle of the polarized light is given by these two readings. From this angle one may estimate the refractive index, absorption coefficient and average thickness of the adsorbed layer. The ellipsometer is limited to one wavelength because one must use a different quater wave plate for different wavelengths of light. The complete theory of ellipsometry is very involved and can be found in specialized optics texts.

Ellipsometry is very sensitive to changes in the surface layer thickness and can detect layers down to 1A average thickness. Thus it gives valuable information on any adsorbed layers on electrodes. However, its usefulness to the study of unadsorbed intermediates of electrode reactions is very limited.

The optical methods described have been in use only for the last few years. Their potentialities have not been fully explored and therefore, general assessment is impossible. Studies using these methods have usually been confined to systems that are fairly well understood from other studies in order to learn these new techniques and evaluate them. Naturally they are the subject of intensive research efforts.

Appendix
Operational Amplifiers

The operational amplifiers are high gain amplifiers used in conjunction with a variety of negative-feedback networks, usually to perform the electrical analogue of mathematical operations on the input signal. A general configuration is given

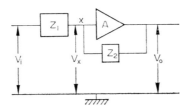

FIG. 82. A general scheme for an operational amplifier.

in Fig. 82. The potential V_x is given by

$$V_x = \frac{Z_2}{Z_1+Z_2} V_i + \frac{Z_1}{Z_1+Z_2} V_0. \qquad \text{A.1}$$

The input to the amplifier at point x is not V_i but

$$\frac{Z_1}{Z_1+Z_2} V_i.$$

The output is not $V_i A$ where A is the amplification, but

$$\left(1 - \frac{Z_1}{Z_1+Z_2} A\right) V_i$$

because part of the voltage is fed back to the input. Thus, the real amplification, $A' = V_0/V_i$ is

$$\frac{V_0}{V_i} = \frac{\dfrac{Z_2}{Z_1+Z_2} A}{1 - \dfrac{Z_1}{Z_1+Z_2} A} = -\frac{Z_2}{Z_1} \frac{1}{1 - \dfrac{1}{A}\left(1 + \dfrac{Z_2}{Z_1}\right)} \qquad \text{A.2}$$

if $A \gg (Z_2/Z_1)$, A.2 becomes

$$V_0/V_i \simeq -Z_2/Z_1. \qquad \text{A.3}$$

This equation is usually correct because Z_1 and Z_2 can be chosen to comply with the condition stated above. If one introduces equation A.3 to A.1, one gets

$$V_x = \frac{Z_2}{Z_1+Z_2} V_i + \frac{Z_1}{Z_1+Z_2} \left(- \frac{Z_2}{Z_1}\right) V_i = 0. \qquad \text{A.4}$$

This means that point x is kept at ground potential by the amplifier and is, therefore, called "virtual ground".

In potentiostats P is equivalent to Z_1, and the capacitor D is equivalent to Z_2. We are not interested in any output, only in the maintenance of ground potential at point x. Therefore the whole output of the amplifier is used as feedback.

Operational amplifiers are used to carry out a multitude of mathematical operations. For information on these various uses one is referred to the literature.†

† A good starting point in the study of electronics in chemistry and electrochemistry is E. J. Bair, *Introduction to Chemical Instrumentation*, McGraw–Hill Book Co. Inc. (1962).

6. Some Technological Aspects of Electrodics

The first five chapters of this monograph presented the theory and described some experimental methods of electrodics. This chapter will present brief descriptions of several technical applications. Five topics will be dealt with here and these have been chosen to illustrate two points: (1) the way by which the theory of electrodics can be profitably used for the solution of some problems (as in analysis or corrosion, for example) and (2) the problems which arise to limit the applicability of electrodic devices (e.g. in fuel cells). In the space of one chapter it is impossible to discuss all the applications of electrodics in industry in any detail, this chapter is therefore limited to very brief descriptions of the principles of the various applications, omitting technical details.

In many technical applications of electrodics, practice has preceded theory by many decades. For example, electrodeposition of metals was practised in the last century, many years before the work of Tafel or Butler and Volmer, who laid the foundation of electrode kinetics. Indeed this field has only recently become the subject of intensive basic research. On the other hand, the application of electrodics to analytical chemistry follows the theoretical studies very closely, mainly due to the fact that the same instruments are used for basic research and analytical research and application with very little engineering being involved. In the field of fuel cells, the theoretical knowledge exceeds the practical since the problems in this area are largely non-electrochemical.

A. Electrodic methods of analysis

The importance of analytical chemistry in all areas of chemistry cannot be overstated. An analytical laboratory is a part of any chemical plant and any research institution. Analytical chemistry employs a wide variety of techniques for its purposes and almost all research techniques, including electrodics, have found application in it. When comparing the electrodic techniques to spectrophotometric techniques (which are the most widely used), the electrodic techniques, in general, need less expensive equipment. However, the electrodic methods of analysis are usually more time consuming than the spectrophotometric ones.

Electrodic methods of analysis can be divided into four main groups according to the property being measured. (1) voltage, (2) current, (3) current-voltage curves and (4) quantity of electricity. This section will, therefore, be divided to four respective sub-sections.

(1) *Methods Based on the Measurement of Voltage (Potentiometry)*

These methods are divided into potentiometry with zero current (classical potentiometry) and potentiometry with current flow.

(*a*) Potentiometry with zero current is a very well known technique in analytical chemistry. The theory involved and the experimental methods used are treated in many standard texts on analytical chemistry and instrumental analysis. Assuming that the reader is familiar with pH measurement with the glass electrode and the various potentiometric titration methods (neutralization, redox, precipitation and complexation) let us examine two less well known aspects of the field: direct potentiometry and potentiometric titration in non-aqueous solvents.

Direct potentiometry is the determination of the concentration of a solution constituent by measuring the voltage of a cell in that solution. In order to do that one needs an electrode reversible to the species to be determined, a reference electrode and an accurately known function which relates the measured voltage to the concentration. When the substance to be determined is hydrogen ions, we have several reversible electrodes, the most convenient of which is the glass electrode. The reference electrode is usually the saturated calomel electrode. The Nernst equation relates the measured cell voltage to the activity of the potential determining species, but not directly to concentration; the difficulties of relating activities and concentrations in a general way are very well known. This problem can be overcome by very careful measurement of the pH of several standard buffers, covering

TABLE X. *Values of* pH *of standard buffer solutions at* 25°C

Standard solution	pH	Name of compound
0·05 M KH$_3$(C$_2$O$_4$)$_2$.2H$_2$O	1·68	Potassium tetroxalate
KHC$_4$H$_4$O$_6$ (saturated solution at 25°)	3·56	Potassium hydrogen tartarate
0·05 M KHC$_8$H$_4$O$_4$	4·01	Potassium hydrogen phthalate
Mixture of 0·025 M KH$_2$PO$_4$ and 0·025 M Na$_2$HPO$_4$	6·86	Potassium dihydrogen phosphate and disodium hydrogen phosphate
0·01 M Na$_2$B$_4$O$_7$.1OH$_2$O	9·18	Borax
Ca (OH)$_2$ (saturated at 25°)	12·45	Calcium hydroxide

the whole pH range at temperatures between 0° and 95°C. Table X gives values of the pH of the standard buffers at 25°C. Thus, the pH meter is calibrated using one of the standard solutions, the pH of which is closest to that of the unknown solution and the pH of the unknown is then measured. The pH cannot, of course, be measured more precisely than to $0 \cdot 01$ unit because the standard solutions are not calibrated more accurately.

The development of the glass electrode, which is reversible to hydrogen ions, was a major contribution toward the development of the measurement of pH. A common glass electrode consists of a thin bulb of a special glass (made of silica and sodium and calcium oxides), responsive to hydrogen ions; this construction makes the glass electrode a high impedance device. The bulb is filled with a solution of constant pH and an inner reference electrode, which is reversible to the anion in the constant pH solution, is placed in it. A common such solution is hydrochloric acid and the inner reference electrode is silver chloride. A schematic picture of the electrode is shown in Fig. 83.

FIG. 83. A schematic picture of a glass electrode; (a) A thin bulb of special glass; (b) A holder of regular, thick glass; (c) The level of the HCl solution; (d) The AgCl/Ag reference electrode.

When aluminium oxide is added to the other ingredients of the glass for glass electrodes, the sensitivity to hydrogen ions diminishes considerably in neutral and basic solutions. Instead, sensitivity toward the alkali metal ions becomes considerable. This phenomenon was used to manufacture sodium and potassium sensitive glass electrodes. Clearly, the possibility of measuring the concentrations of metal cations with the same ease that one measures the pH, has excited many in the field of glass electrode research. Many glass electrodes are now known which are sensitive to cations other than sodium and potassium, but the usefulness of these electrodes has not yet been fully exploited. The reason is that cation sensitive glass electrodes are sensitive to several cations, the formula giving the potential of this electrode in a solution containing two cations is not the simple Nernst equation but

$$V = \text{constant} + \frac{nRT}{F} \ln [a_i^{1/n} + (K_{ij}^{\text{pot}} a_j)^{1/n}] \qquad 6.1$$

where V is the potential of a cell containing the glass and a reference electrode, i and j are the two cations in solution. The constant is determined by the nature of i and j and by the reference electrode used. a_i and a_j are the activities of i and j respectively, K_{ij}^{pot} is the relative sensitivity of the glass electrode toward i relative to j and n is a parameter, specific to i, j and the glass used. Thus, the problem to be solved before glass electrodes can be used for the determination of any cation, is to find the glass for which the K^{pot} value will be large for this cation with respect to all others.

Once suitable glass electrodes are found, Tables such as Table X will have to be determined and the potential values, read on the "cation meter", will have to be correlated in an unambiguous way with the concentrations of the cations in solution. This is essentially a standardization procedure, one which will be very fruitful in many ways and will, no doubt, follow the development of useful glass electrodes.

The second subject to be considered is that of potentiometric titration in non-aqueous solvents. When we titrate A with C the reaction

$$A + C = B + D \qquad\qquad I$$

has the equilibrium constant K. It is obvious that the greater is K, the easier it is to measure the end point and the closer is the measured end point to the true equivalence point. In other words, when K is large all of A is transformed to B when a very small excess of C is present in the titration mixture. If K is small, a very large excess of C is needed to convert all of A to B; a meaningful titration for reactions with small equilibrium constants is not possible. Consider, for example, the titration of pyridine with acid in water

$$C_5H_5N + H_3O^+ = C_5H_5NH^+ + H_2O \qquad\qquad II$$

The equilibrium constant of reaction II is not as large as needed for the satisfactory determination of pyridine in water, i.e. a considerable excess of acid is needed in order to convert all the pyridine to pyridinium ion. If, however, we react pyridine with an acid much stronger than H_3O^+, i.e. an acid that transfers its H^+ ions more readily, we should be able to convert all the pyridine to pyridinium ion much closer to the equivalence point. It is clear that any such acid cannot be used in water, since all acids in water form H_3O^+. In glacial acetic acid, the reaction will be

$$C_5H_5N + CH_3COOH_2^+ = C_5H_5NH^+ + CH_3COOH \qquad\qquad III$$

and since $CH_3COOH_2^+$ is a much stronger acid than H_3O^+, reaction III will have a very high equilibrium constant, i.e. it goes to completion at the equivalence point and can be used for the determination of pyridine by direct potentiometric titration. Titrations in non-aqueous solvent are necessary whenever the titration reaction I does not go to completion in water or when the reagents are insoluble or unstable in water. In principle, there are no special problems associated with non-aqueous titrations. The problems which are encountered are technical in nature. One of these is the construction of a suitable reference electrode for the solvent. Sometimes an aqueous reference electrode (such as the aqueous s.c.e.) is used, but then one must make sure that the liquid junction potential between the aqueous and the non-aqueous parts is reproducible. Indicating electrodes in non-aqueous systems may also behave in a different way than in aqueous ones, e.g. the glass electrode was found to be sluggish and unsuitable for acid base titrations in many solvents except in glacial acetic acid, methanol, acetonitrile and dimethylformamide.

The use of non-aqueous potentiometric titrations has been rather limited, perhaps due to lack of knowledge of chemical reactions in these solvents. Better understanding of chemistry and electrochemistry in non-aqueous solvents will, no doubt, result in wider use of these interesting titrations.

(b) Potentiometry with current flow uses either one indicating and one reference electrode in the measuring cell, or it may use two indicating and no reference electrode in the cell. The latter is illustrated in Figs 84 and 85.

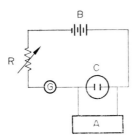

FIG. 84. The circuit of potentiometry with constant current; A: a voltage measuring instrument; B: batteries; C: cell; R: resistor; G: galvanometer.

Figure 84 shows the measuring circuit used for this method and Fig. 85 shows the current-potential curves for the titration reaction I. When a small current I is passed through the cell, this current must pass through the cathode, producing a reduction reaction and through the anode, producing an oxidation reaction. At the beginning of the titration only A is present in the cell, the anodic reaction will be the oxidation of A to B and the cathodic reaction will be the reduction of water and evolution of hydrogen. The potential of the cell will be V_i. When the titration is half completed species A, B and D are present in the solution. The anodic reaction will be, as before, oxidation of A, but the cathodic reaction will be reduction of B and the potential of the cell will change from V_i to V_m. At the end point the solution contains only species B and D. The cathodic reaction will not change ($B = A$), but the anodic reaction will now be the oxidation of D and the voltage of the cell will be V_e. After excess of C has been added to the cell, the cathodic reaction will change to the reduction of C and the anodic reaction will remain the oxidation of D, the voltage will now be V_x. A plot of the voltage

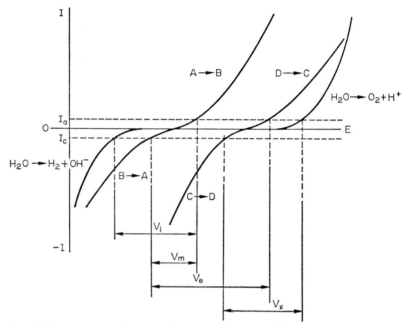

FIG. 85. Current potential curves for potentiometric titration with constant current.

against volume of titrant added will look like the curve of Fig. 86, the voltage is small except at the end point where there is a sudden increase. This increase can be used to control a relay to stop the titration, automating it.

The shape of the voltage against volume of added titrant curve depends on the current potential characteristics of the substances involved in the

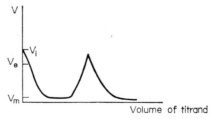

FIG. 86. Voltage against volume of titrant curve for potentiometric titration with constant current.

titration. Thus one must know these current-voltage curves in the system to be titrated. This is not easy when the titration mixture is very complex, as it often is. The need to predict the shape of cell-voltage against volume of added titrant curve for every system separately makes the general evaluation of the method of potentiometric titration with current flow very difficult if not impossible. The method is not of very wide practical use, its interesting features are its simplicity and the field of unexplored possibilities which it opens to the interested.

(2) *Methods Based on the Measurement of Current (Amperometry)*

The linear dependence on concentration of the limiting current at the d.m.e. or at solid electrodes in stirred solutions is the basis of the method known as "amperometric titration". The titrant or the unknown must be electroactive. The potential of a simple polarographic two-electrode cell is fixed at a value in the limiting current region of the electroactive species. If the unknown is electroactive, the current is initially high, but drops steadily as the unknown is consumed by the titration reaction and, if the titrant is not electroactive, reaches zero at the end point. If the titrant is electroactive, the current stays zero until the end point is reached, when it starts to rise steadily. If both the titrant and the unknown are electroactive, the current first decreases and then increases again beyond the end point to give a V shaped curve. The shape of the titration curves are shown in Fig. 87. An interesting application of amperometric titration is that sometimes the titrant need not react with the unknown. If the titrant is reduced at the same potential where the unknown is oxidized or vice versa and if the diffusion coefficients of the two substances are similar, the addition of the titrant will depress the current

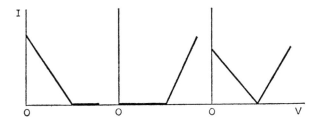

FIG. 87. Typical curves of amperometric titrations.

due to the unknown at the d.m.e. When the resultant current is zero i.e. the cathodic current is equal to the anodic current at the d.m.e., the concentration of the unknown is equal to the concentration of the titrant and can be calculated.

The advantages of amperometric titrations over other titration methods are twofold. First, the reaction need not go to completion at the end point since the latter is determined from the intersection of two straight lines (similar to conductometric titrations but different from potentiometric titrations). Second, the concentration range of the titrated solutions is $10^{-2}-10^{-4}$ Molar, a concentration range where other titration methods usually fail. The disadvantage of amperometric titration is the time it consumes. Each point has to be determined separately by adding the titrant, stirring the solution and reading. Amperometric titrations are widely applied in analytical chemistry.

A special class of amperometric titrations is known by the rather odd name of "dead stop end-point determination". The cell in this method includes two platinum electrodes (no reference electrode) between which a small voltage is applied. The best known determination which uses this method is the water determination by the Karl–Fischer reagent. The reagent is a complex of iodine, sulphur dioxide and pyridine; its reaction with water converts the iodine to iodide. Thus, before the end point is reached, the solution contains iodide, which can be oxidized at the small applied voltage, and nothing that can be reduced. Thus, there is no current flow through the cell (in order that direct current may flow through a cell, two electrode reactions have to take place—oxidation at the anode and reduction at the cathode). When the first drop of excess iodine is added with excess Karl–Fischer reagent, iodine can be reduced at the cathode and current starts to flow through the cell. The point where the current increases sharply is the end-point of the titration. This method for end-point determination can also be automated, but its application is limited to titrations involving reversible redox couples.

(3) *Methods Based on the Determination of Current-Voltage Curves*

Polarography is one of the most widely applied methods in analytical chemistry. Most of the metal cations, many anions and several organic compounds are determined polarographically in a variety of mixtures. A schematic drawing of a simple polarographic circuit was given in Fig. 48. The cell is usually a two electrode cell, containing a working and a reference electrode. The d.m.e. is by far the most popular working electrode, but rotating platinum electrodes are also widely used. The reference electrode is usually the s.c.e. When the solution to be analysed forms precipitates with K^+ or Cl^- ions, electrolytes other than KCl are used in the salt bridge between the electrode compartments. The concentration range of polarography is $0 \cdot 01 - 0 \cdot 0001$ M but in some favourable cases, concentrations of 10^{-5} M have been determined.

The cell solution should be conducting, i.e. it should contain a fairly high concentration of non-electroactive electrolyte. The solution may contain several electroactive species, if their waves are well separated and the concentration of any one species does not exceed a few millimolar, the determination of one substance is not affected by the presence of the others. If the wave to be determined is very close to another wave, measures must be taken to eliminate the interfering wave.

The advantages of polarography are that the presence of foreign non-electroactive electrolytes is not an objection, but actually an asset, the method is fairly accurate, precision of 1% in many cases being obtained. Often mixtures need not be separated before analysis and determination of two or three substances in the same solution can be made simultaneously. The most outstanding disadvantage of polarography is the time required per determination—an average one may take as long as 20 minutes. When polarography is used as a routine analytical method it is worth while to shorten this period by the use of several cells, each one at a different stage of the determination.

The essential part of polarography is the determination of the current-voltage curve, and measurement of the half wave potential and the limiting current. The former helps to identify the unknown and the latter is proportional to its concentration. Figure 88 shows the way in which these parameters are measured; the best method is the one which gives exact proportionality between the limiting current and the concentration. It must be remembered that if the limiting current is diffusion controlled (which is

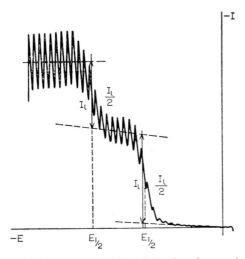

FIG. 88. Measurement of I_1 and $E_{1/2}$ in polarography.

the case for most substances) then it must be proportional to concentration, if it is not, the fault is in the method of measurement and not in the system. Therefore, provided that the limiting current is diffusion controlled, the best method of measurement is the one which results in a straight line relationship between i_1 and concentration. Figure 88 also shows how to determine the limiting current and half wave potential for the second and other waves which in polarography is the same as for the first wave. However, in solid

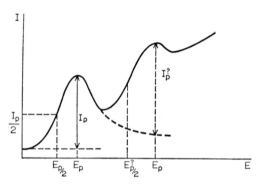

FIG. 89. Measurement of I_p, E_p and $E_{p/2}$ in voltammetry.

electrode voltammetry in quiescent solutions, the determination of the peak current of the first wave is straightforward, but that of the second and other waves is very complicated (see Fig. 89). This complication limits the usefulness of voltammetry in quiescent solutions for analytical purposes; voltammetry in stirred solutions is free from this complication.

All polarographic experiments suffer from the interference of oxygen and therefore, it must be eliminated by bubbling nitrogen through the solution. Voltammetry is usually done in potentials positive to that of the reduction of oxygen and, therefore, does not interfere in this potential range.

The appearance of maxima may interfere with polarographic determinations; maximum suppressors may be used (cautiously) to improve the results. Again, voltammetry is free from this interference (see Chapter 5, Section D and Fig. 61).

The calibration of the polarographic data, i.e. finding the proportionality constant between i_1 and concentration, can be done in one of three ways. The first method is to prepare a series of standard solutions, measure the limiting current and construct a calibration curve which should look like the one in Fig. 90. This method is best for establishing a measuring technique and is very easy to use when one has to perform a series of determinations on the same substance. If a single determination only is to be made, the better

method is the pilot ion method. Here an ion is chosen, different to the one
to be determined, whose wave is well separated from the wave to be determined, as the pilot ion. One calibration experiment is needed, where the

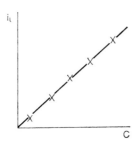

FIG. 90. A typical polarographic calibration curve.

concentrations of the pilot ion and the ion to be determined are known.
Then, from Ilkovic equation (equation 3.60)

$$\frac{I_{11}}{I_{12}} = \text{const.}\,\frac{c_1}{c_2} \qquad\qquad 6.2$$

where I_{11} and I_{12} are the limiting currents of the ion to be determined and
the pilot ion, respectively; c_1 and c_2 are their respective concentrations in the
standard solution. The constant can be calculated directly. A known concentration of the pilot ion is added to the unknown solution and the current
voltage curve is recorded. It is then a simple matter to determine the concentration of the unknown from the following equation

$$c_x = \frac{I_{1x}}{I_{12x}}\,\frac{c_{2x}}{\text{const}} \qquad\qquad 6.3$$

where I_{1x} and c_x are the limiting current and concentration of the unknown
in the solution under analysis; I_{12x} and c_{2x} are the limiting current and concentration of the pilot ion in the analysis solution. A check on the measurement is given by the limiting current of the pilot ion: this should be proportional to the concentration of the pilot ion in the calibration and the measurement curves. If the operator is experienced in measurement of limiting
currents, he can use the method of standard addition for measurement of
concentrations. The procedure here is to take the polarogram of the unknown
solution, add a known concentration of the same substance and record the
current-voltage curve again. If it is assumed that the measured limiting
current is proportional to concentration, the concentration of the unknown

is given by

$$c_x = \frac{c_s I_{1x}}{I_{1(x+s)} - I_{1x}}$$ 6.4

c_x is the unknown concentration; c_s is the concentration of the standard addition; I_{1x} is the limiting current of the unknown and $I_{1(s+x)}$ is the limiting current of the unknown and the standard addition.

When two substances are reduced at the d.m.e. at potentials which are very close to each other (as in Fig. 91) measurement of limiting current may be impossible. There are several ways of tackling this problem one of which is

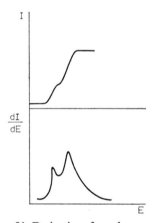

FIG. 91. Derivative of a polarogram.

to take a derivative of the polarogram. Most modern polarographs include a derivative circuit. The derivative of the polarogram looks like a peak-shaped curve, the peak is at the half wave potential and the height of the peak is proportional to concentration. Thus, many waves which are too close to be measured directly could be dealt with using the derivative curve (see Fig. 91).

The supporting electrolyte has a strong influence on the half wave potential; its influence is particularly strong if the ion studied forms a complex with the supporting electrolyte (or any other constituent in the solution). In general the half wave potential of a metal complex is more negative than the half-wave potential of the aquo ion. The larger the stability constant of the complex, the larger is the shift of potential. Moreover, the half wave potential also changes with the concentration of the complexing agent; higher concentration of the complex-forming substance results in a larger shift of the half wave potential to the negative. Thus, when the waves of two ions are too close to be measured, the addition of a complex forming agent can modify the curve by making one wave much more negative than the other.

Sometimes the solution to be analysed contains a large concentration of an easily reduced substance and a small concentration of another substance, which can be reduced only at a rather negative potential. The polarogram of such a combination is given in Fig. 92. It is clear that the second sub-

FIG. 92. A small wave follows a large wave.

stance cannot be determined without separating out the first. A convenient way of separation is electrolysis of the solution at a constant potential, so that all of the first substance is completely reduced; the second material can now be determined easily.

When the limiting current is the same as the residual current at the same potential, the limit of sensitivity of polarography is reached. To minimize the residual current is the best way to increase the sensitivity and, hence, the scope of polarography. The residual current results from two processes: (i) the charging of the double layer of the electrode [this charging current is especially large with the d.m.e. because of the change of the area of the electrode with time (see equation 2.3)] and (ii) electroactive impurities which are present in the solution, their electrolysis current being added to the charging current. Naturally, we try to work with clean solutions, but there is a limit to the extent to which one can purify solutions in an analytical laboratory; too extensive purification results in lost material and time. Fortunately, the residual current in similar solutions is similar under normal circumstances. Thus, if we put two polarographic cells with two d.m.e. so that the current of one is subtracted from the other, we may eliminate the residual current. In order to do so, the drops of the electrodes must be synchronized so that they will have the same area at the same time. This is done by using a mechanical beater that knocks the drops off the capillaries before they fall naturally. Thus, when one of the cells contains the unknown and the other does not, the "signal-to-noise ratio" of the polarograph is improved very much and the lower limit for determination of substances by polarography is reduced to 10^{-6} M.

When the substance to be determined is not electroactive, indirect methods may be used to determine it. One such method is to react the inactive compound to form an active one, e.g. glycols may be oxidized to aldehydes— glycols are polarographically inactive, but aldehydes are reducible. Another method involves the reaction of the unknown with an excess of active compound, polarographic determination of the excess and calculation of the original concentration. Many other methods, known in analytical chemistry, can be used for indirect polarography to advantage. The disadvantages of these methods are those of indirect analyses—the reaction which converts inactive substances to active ones must be quantitative under the conditions of the analysis. Thus, one would use indirect analysis only when this is the best or only way of determining the unknown. An illustrative example for an analysis that is best done by an indirect method is the determination of sulphate in dilute solutions. This is done by precipitation of the sulphate ions as lead sulphate and dissolution of the precipitate in ammonium acetate; the lead is then determined polarographically.

(4) *Methods Based on the Determination of Amount of Charge* (*Coulometry*)

These methods are divided into two parts: the first is controlled potential coulometry and the second is coulometric titration. The principles of the two methods are quite different, the only common features are that the quantity measured is the amount of electricity used and the concentration is calculated using Faraday's law.

(*a*) Coulometry at constant potential. The possibility of controlling the potential of the working electrode by the use of potentiostats was discussed in Chapter 5, Section A. If the potential is controlled at some point on the limiting current portion of the current potential curve, we can electrolyse all the material present in the solution. If the circuit includes a coulometer, which can be either a silver coulometer or an electronic current integrator, we can measure the total amount of electricity consumed in the electrolysis. This amount of electricity is proportional to the amount of material electrolysed by Faraday's law; the concentration is then determined by dividing the amount of material by the volume of the solution. The current at any one time is proportional to the concentration of the electroactive substance (in stirred solutions)

$$I_1 = nFADc_0/\delta \qquad\qquad 6.5$$

Faraday's law states

$$Q = nFW/M \qquad\qquad 6.6$$

where Q is the amount of electricity consumed, W is the weight of material electrolysed and M is its molecular weight. In differential form

$$I = \frac{dQ}{dt} = \frac{n\mathrm{F}}{M}\frac{dW}{dt} = n\mathrm{F}V\frac{dc}{dt} \qquad 6.7$$

V is the volume of the solution. Equating 6.5 and 6.7 and integrating yields

$$c = c(0)\exp -\frac{AD}{V\delta}t \qquad 6.8$$

$c(0)$ is the concentration at the beginning of the experiment, $t = 0$. Equation 6.8 shows that the current decreases exponentially with time, the rate of decrease does not depend on the concentration, but depends on the electrode area and on the rate of stirring (via δ). Thus, theoretically the current would never decrease to zero and the electrolysis would never come to an end, but in practice this means that one can electrolyse as much of the material as the accuracy of the determination demands. For general use, electrolysis to 0.1% of the original current value is enough, but for special purposes one may want a more accurate determination.

The above discussion assumes that all the current is provided by the measured electrode reaction and no current is used for any other process. This assumption is justified if the residual current is very much smaller than the electrolysis current, i.e. the species that is being electrolysed is the only electroactive substance in the solution at that potential. Obviously, if there is more than one substance in the solution which is electroactive in the applied potential, the amount of electricity consumed is the sum of the amounts consumed by all substances and is, therefore, related to the sum of their concentrations. The current efficiency is defined as that fraction of the current due to the electrode reaction of interest and is usually given in per cent

$$\text{current efficiency} = \frac{\text{current due to process of interest}}{\text{total current}} \times 100$$

Establishment of the current efficiency is a prior requirement to any attempt for the development of a coulometric method of analysis.

Coulometry at a constant potential is the most accurate method of analysis known. Quantities of electricity can be measured to a precision of six or more significant figures; deposits of metals can be weighed to an accuracy of a small fraction of a miligram. Thus coulometry at a constant potential serves as a standardizing method for other methods of analysis.

(b) Coulometric titration, as its name implies, is basically a titration. The cell consists of two electrodes and it does not matter if neither of them is a good reference electrode because potentials are not measured or controlled in this technique. The solution contains, beside the unknown and supporting electrolyte, the reagent precursor, i.e. a substance which is oxidized or reduced to give the reagent with which the unknown reacts. For example, the amount of double bonds in petrol can be determined by passing a constant current between two platinum electrodes dipped in a solution of the petrol and $1M\ KBr$ in a mixed solvent. The bromide is being oxidized at the anode to bromine which reacts with the double bonds. The end point must be detected by a separate method. Since the titration is carried out at constant current, the amount of electricity is given by the current intensity multiplied by the time of the titration. This amount of electricity is proportional, by Faraday's law, to the amount of reagent consumed and thus to the amount of reactive material (in our example, the "amount" of double bonds) in the solution. The concentration is again found by dividing the amount by the volume of the solution.

Sometimes the electrode itself is the reagent precursor; for example the titration of chloride in water is done by using a cell the anode of which is silver and the cathode is platinum. Silver ions form at the surface of the anode and precipitate the chloride ions, the cathodic reaction is the evolution of hydrogen. Again the end-point is determined by a separate method.

Coulometric titrations are very useful since one can titrate unknowns with unstable reagents, such as radicals or other intermediates. A further advantage is the absence of the need to prepare and standardize solutions. The disadvantage of coulometric titrations is that they are not generally applicable, every system has to be developed and evaluated independently.

B. Electroplating

Electroplating of metals is used extensively for very many purposes: protection against corrosion by nickel or gold plating, decorative plating, e.g. bright chrome on cars, and protection against mechanical erosion as in hard chromium plating of many moving parts, are just a few of the better known uses. This section will first discuss the properties of the deposit, their

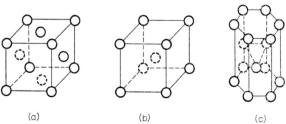

(a)   (b)   (c)

FIG. 93. (a) Face centred cubic crystal; (b) Body centred cubic crystal; (c) Hexagonal closed packed crystal.

correlation with its structure and the structure of the bulk metal. Then some general considerations for devising a plating process will be given, followed by description of plating baths, their constituents and maintenance. Finally, a few words will be said on the basic research connected with electroplating.

The metallic electrodeposit is crystalline in the same sense that thermally prepared metals are crystalline, i.e. it consists of arrays of atoms arranged in specific order, repeated along the three co-ordinate axes. Small crystals that touch each other at the surface are known as grains. Three main crystal structures are known in metals: the face centred cubic, body centred cubic and hexagonal closed packed (Fig. 93). Ductility is associated with the face centred cubic structure and brittle metals are hexagonal.

The structure of the grain boundary, like that of any surface, differs from the structure of the bulk. Therefore, the properties of grain boundaries differ from those of the bulk. The result is that when the grains are large, the amount of material in the boundary regions is small compared to the total amount of material and the properties of the metal are influenced mainly by its crystal structure. When the grains are very small, the amount of material in the boundary region is large compared with the total amount of material and the properties of the metal as a whole will be influenced more by the surface properties of the grains than by the crystal structure of the metal. Since most commercial deposits have very small grain sizes, it is expected that the metallurgical properties of the deposit, such as hardness, ductility and strength, will depend less on the crystal structure of the metal and more on factors which determine grain size, amount of adsorbed and codeposited foreign material and other such surface properties. In practice, the properties of small grain deposits may vary greatly from those of the same metal prepared conventionally. Good examples can be obtained from alloy deposition; often, compounds and solid solutions unobtainable by thermal methods are observed in electrodeposits, e.g. the alloy of $\frac{1}{4}$ tungsten and $\frac{3}{4}$ cobalt is electrodeposited as a solid solution, but exists only as separate phases in thermally prepared samples under 1000°C.

Experience of platers showed that, as a rule, large grain deposits tend to be soft, ductile and weak, while small grain deposits are usually hard, brittle and strong. However, this rule, like very many others in electroplating, does not always apply, one could find specific cases when the reverse of this rule applies. Moreover, the properties of the deposits cannot be correlated with grain size inside the groups of "large grains" and "small grains", especially in the last group. Properties such as brightness or strength depend more on the bath constituents than on the grain size.

The testing of electrodeposits for hardness, strength and other properties presents a problem to the metallurgist. Often the deposit must be stripped off for testing. This subject, however, will not be discussed here.

When electrolysing any solution, a reaction takes place only when the cell voltage exceeds a certain value typical of the solution. When the voltage is increased, the current increases but eventually a limiting value is almost always reached. Mass transport controlled limiting currents are observed in concentrated solutions as well as in dilute ones. Admittedly, this limiting current is very much larger in the former case. There is an interesting relationship between the grain size of the deposit and the current density; at small current densities the grains are large and often distinct single crystals are formed on the surface. At medium high current densities the deposit is of fine grains and may be referred to as "polycrystalline"; at very high current densities, often at the limiting current, the deposit becomes very loose and looks "burnt". Obviously it is useless for practical purposes.

The cathode in a plating bath is the article to be plated, the other constituents are determined by consideration of quality and cost. The anode of the electroplating cell can be made either of the plating metal (called the "soluble anode") or from some inert material, such as graphite (the "insoluble anode"). The advantage of using a soluble anode is that the concentration of ions in the bath remains constant; when an ion is reduced at the cathode, a similar ion is formed at the anode. This removes the necessity to monitor the composition of the bath solution as rigorously as when the metal cations are removed from the solution and not replaced. However, in some cases, the best example of which is chromium plating, a soluble anode cannot be used. A chromium anode will dissolve to form Cr^{3+} ions, but the chromium species from which the plating is obtained is $Cr_2O_7^{2-}$. Other cases where a soluble anode cannot be used is when the anode forms an insoluble film on its surface, this film increases the resistance of the cell and therefore its formation is always undesirable.

One of the most commonly interfering processes with electroplating is the evolution of hydrogen. The overpotential of hydrogen evolution varies enormously from one cathode metal to the other from zero on platinum under standard conditions to two volts on mercury; it also varies with the pH of the solution, in alkaline solutions this potential is more negative than in acid.

It is interesting to note that some metals on which the overpotential for hydrogen evolution is small (i_0 is large), have a high overvoltage for deposition of this metal (i_0 is small) and vice versa. The most striking example (admittedly not relevant to electroplating) is that of Hg. The overpotential for hydrogen evolution is the largest of all metals and the overpotential for the reduction of mercurous salts is one of the smallest known.

Plating baths may contain all or part of the following components: the metal compound to be electrolysed and inert electrolyte to increase conductivity, a compound that promotes the anodic reaction (this is either a reducing agent, to be oxidized at the anode or a material which dissolves films formed on the anode), buffer which keeps the pH constant and "brighteners" or "addition agents" which help to produce a smooth, bright deposit. When gas (hydrogen) is formed at the cathode during the electro-plating process the bath may also contain a wetting agent which speeds the discharge of bubbles from the cathode, thus preventing the formation of pits in the deposit.

In electroplating, as in all branches of technology, the engineer has to strike a balance between the quality of the deposit and the cost of obtaining it. The cost is composed of capital cost (the price of buildings, permanent equipment etc.), running cost (the price of materials and power) and labour cost. Thus, the ideal would be to produce a high quality deposit using cheaply built cells and inexpensive materials, the process requiring little more than the revisible potential. One can approach this ideal by using the cheapest materials which give an acceptable deposit and design the cell and the bath constituents in such a way as to minimize the voltage required. Another factor which must be considered when calculating the costs of the process is the current efficiency.

Figure 94 shows a typical relationship between the current-potential curve of metal deposition and that of hydrogen evolution. It shows that at the high current density used for electroplating, the potential is more negative

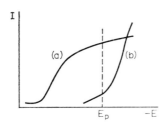

FIG. 94. Current voltage curves showing a small current efficiency in plating; (a) Metal reduction; (b) Hydrogen evolution. E_p potential of plating.

than that of hydrogen evolution. Thus, the current efficiency of very many plating processes is lower than 100%. The lower the current efficiency, the more expensive the process. One of the factors which determine the pH of the bath is the hydrogen evolution potential and, consequently, the current efficiency.

Plating baths are divided into "acid" and "alkaline". Acid baths contain simple salts of the reducible metal which are often chosen according to their solubility, the purpose being to obtain a solution which is as concentrated as possible. The high concentration is needed to minimize the solution resistance and to increase the current density at the cathode. Nitrates are all soluble, but they are not generally used because of the reducibility of the nitrate anion. Chlorides are very useful, and so are sulphates, although it should be borne in mind that chlorides may form undesirable complexes with the metal cation to be plated. Other anions of more limited use in plating baths are the perchlorate and fluoborate. The use of the former is limited because of the risk of explosion and the latter is quite expensive. Lead, however, is often plated from fluoborate solutions because both the chloride and the sulphate are of limited solubility. Some of the simple salts undergo hydrolysis in neutral pH. Therefore acid is often added to the bath to depress this hydrolysis, hence the name "acid baths".

Alkaline baths contain complex compounds of the metal to be plated and are, as their name implies, alkaline. The most common ligand used is cyanide (for the plating of copper, cadmium, gold, silver and zinc, for example) and since it hydrolyses at neutral pH, sodium or potassium hydroxide or excess cyanide is added to the bath to repress this hydrolysis. Alkaline baths absorb carbon dioxide from the atmosphere to form carbonates. When the carbonates precipitate the metal to be plated this can be a serious problem. A general solution, however, is not available.

Some of the baths which contain complexes of the deposited metal are not alkaline and do not contain cyanide, but other ligands. Chromium is plated from a solution of chromic acid, tin is likewise plated from a solution of its oxy-complex. Chloro complexes are also sometimes used for gold plating and many other complexes are used for special purposes.

The word "buffer" is used in the electroplating literature rather loosely, sometimes the compound called "buffer" really functions to keep the pH constant (thus check on current efficiency, for example) but at other times this word is used to describe a compound which is added to stabilize the solution or promote anode reaction.

The most common brighteners used in plating baths are organic compounds having $-NH$ or $-SH$ bonds. The exact nature of the compounds used is one of the best kept secrets of the plating industry. Addition of brighteners helps to create polycrystalline, smooth deposits over a fairly large range of current density. In plating baths which contain brighteners the value of the current density needed to produce polycrystalline deposits is much lower than that in baths which do not contain these agents. In cases where baths without brighteners do not produce this polycrystalline deposit at all, the addition of the proper compound makes the formation of this deposit possible. It is believed that the addition agents adsorb on the more active sites of the cathode, blocking them and thus promoting deposition on the less active sites. In this way the formation of large grains is prevented.

The distribution of current density on the cathode is not uniform, edges and protrusions carry a higher current density than recessions and dents.

Normally the thickness of the plating will be larger in areas of larger current density. This, of course, is undesirable. A smooth plating is wanted, one which covers imperfections in the article rather than accentuates them. The ability of a solution to give a smooth deposit of uniform thickness is referred to as "throwing power" of the solution—if the deposit is uniform the solution is said to have good throwing power. If surface active agents are added to the baths so that they absorb more readily on the edges and protrusions, the throwing power of the solution is improved. These agents are usually the same ones added as brighteners.

As the bath is worked, the concentrations of its constituents change continuously. The most obvious change is that of the plating metal when an insoluble anode is used. Other changes are the change of pH as a result of hydrogen evolution, a change in the concentration of the addition agents as they become incorporated into the deposit, accumulation of carbonates, dissolution of impurities, etc. Therefore, a frequent check on the composition of the plating bath is needed. Indeed, no plating procedure is complete without suitable methods for the analysis of the bath constituents, ideally without interrupting the plating process. The methods used vary from density measurement (used to monitor the chromate to acid ratio in chromium plating) to sophisticated chemical methods for the analysis of specific components.

Electroplating has been for a very long time in the state of an art rather than an exact science. The fact that the quality of the deposit bears little relation to the structure of the bulk metal or even to the grain size, discourages pure scientific research in this field. Another factor which did not help to arouse the interest of researchers into the basic aspects of electroplating is that a large part of the very wide practical knowledge accumulated by electroplaters over the years is kept secret by companies. This knowledge has produced excellent results at reasonable costs, but has remained outside the realm of the basic scientist. However, many questions remain to be answered. For example: what is the correlation between the metal complex reduced and the structure of the deposit? What is the mechanism by which "addition agents" work? What are the rates of electrodeposition reactions and how are they influenced by bath constitutents? Why does small current produce large grains and large current produce small grains? What is the nature of the species which passes through the double layer to become a deposit? Some of these questions are investigated vigorously, others remain unchallenged.

C. Corrosion

The word "corrosion" is widely used to describe the processes which cause the consumption of metals when exposed to their surroundings, whether in the sea, chemical plant, soil or air. This process is characterized by the oxidation of the metal to form compounds and, as a result, to the reduction of some, as yet unspecified, oxidant. Corrosion is, obviously, a surface process, the laws governing it are the laws of heterogeneous processes and of great importance are diffusion of reactants to the surface and film formation.

In any redox reaction there is a direct relation between the free energy change and the potential for that reaction (equation 1.1), therefore the use of the words "potential" or "free energy" is entirely equivalent. Moreover, any redox reaction can be imagined to take place at electrodes, current potential curves can be drawn for the oxidation of the metal and for the reduction of the oxidant, such as in Fig. 95. The current axis in Fig. 95 is

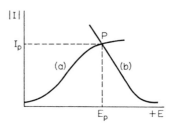

Fig. 95. Current voltage curves showing corrosion parameters; (a) Oxidation; (b) Reduction.

labelled $|I|$ since we disregard the sign of the current for this discussion. In any cell, the cathodic current equals the anodic current, otherwise charge will accumulate in the cell, something which contradicts the law of electroneutrality. At the point P the two curves intersect, the current is I_p and the potential E_p. These are the values of current and potential for the spontaneous redox reaction

$$A + B = \text{corrosion products} \qquad \qquad \text{IV}$$

Conditions which maximize corrosion are those which cause a high I_p: the

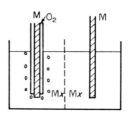

Fig. 96. A differential areation cell in corrosion.

ready availability of oxidant and the presence of a good electrolyte solution in contact with the metal, for example a solution of bromine in KBr, is extremely corrosive; the bromine is a good oxidant and KBr is an electrolyte. On the other hand, pure water is not corrosive, particularly if oxygen is eliminated.

Many oxidants can be responsible for corrosion, but the two commonest are hydrogen ions and molecular oxygen and most of the corrosion occurs in acids or in places exposed to air. Consider, for example, Fig. 96. A cell is shown where both electrodes are of metal M, the cathodic and anodic compartments are separated by a conducting membrane and one compartment is continuously purged with air while the other is kept air free. The cell reaction

$$\text{anode} \quad M = M^+ + e \qquad\qquad\qquad\text{V}$$

$$\text{cathode} \quad 2H^+ + O_2 + 2e = H_2O_2 \qquad\qquad\text{VI}$$

$$H_2O_2 + 2e = 2OH^- \qquad\qquad\qquad\text{VII}$$

is set up and the anode corrodes rapidly. This cell is called "differential areation cell" and is commonly found in corrosion processes.

When the cathodic reaction is the evolution of hydrogen, reduction mechanism may be

$$H^+ + e = H(\text{ads}) \qquad\qquad\qquad\text{VIII}$$

$$H(\text{ads}) + H^+ + e = H_2(\text{gas}) \qquad\qquad\text{IX}$$

or

$$2H(\text{ads}) = H_2(\text{gas}) \qquad\qquad\qquad\text{X}$$

It is evident from the above discussion that the rate of the corrosion process is limited by the rate of the slowest of the anodic or cathodic reaction steps. Thus any inhibition of either metal dissolution, or hydrogen evolution or oxygen reduction will inhibit corrosion.

The metal is oxidized at the anode, therefore only that area acting as an anode will corrode. If the area of the anode is large, the total amount of corrosion (total quantity of electricity involved in the process of corrosion) is distributed over a large area. If, on the other hand, the area of the anode is small, the total amount of corrosion is concentrated in a small area, resulting in formation of pits or even holes in the metal. If the area of the cathode is large and the cathodic reaction is limited by diffusion of oxygen, the total rate of reduction will be high. Thus, a combination of large cathodic and small anodic areas results in very bad corrosion and breakage.

In the following we shall first describe those corrosion reactions accompanied by evolution of hydrogen ; then corrosion in air will be discussed.

This will be followed by sections on passivity and on prevention of corrosion and protection of metals against it.

1. *Corrosion in Acids*

Most metals, when in contact with a non-oxidizing acid, will produce hydrogen by a simple displacement reaction

$$M + nH^+ = M^{n+} + \frac{n}{2} H_2 \qquad\qquad XI$$

The rate of the reaction depends on the exchange current density of hydrogen evolution on M and on the difference between the equilibrium potentials of the hydrogen electrode and the M^{n+}/M electrode. Table XI gives several values of exchange current densities of hydrogen evolution on several metals.

TABLE XI. *Exchange current density* $(A \ m^{-1})$ *for hydrogen evolution on several metals*

Metal	$-\log i_0$
Silver	2·3
Aluminium	6·0
Gold	2·6
Berilium	5·0
Cadmium	8·1
Iron	1·9
Galium	2·7
Mercury	8·1
Nickel	1·4
Lead	8·9
Palladium	−1·0
Platinum	−1·4

The noble metals: Ag, Au, Pd, Pt and Cu are stable in non-oxidizing acid solutions because the free energy of reaction XI is positive for them. Hg and Pb are stable in acid because of the very low rate of hydrogen evolution on them (see Table XI). Another mechanism by which a metal becomes resistant to acid is by formation of a tough, adherent and insoluble film on the surface of the metal. Thus, aluminium, when scraped, dissolves readily in acid, but if a layer of oxide is present on its surface, corrosion is much slower. The use of lead to make sulphuric acid containers is the result of both the low exchange current density of hydrogen evolution on lead and the formation of insoluble lead sulphate. Perhaps the best example for protection by film formation is the use of magnesium vessels to store hydrofluoric acid. Magnesium fluoride is formed on the surface of the metal and prevents any interaction between the metal proper and the acid.

Corrosion in acid is prevented if reaction VIII is slow but if reaction VIII is fast, and reaction IX or X is slow, two things can happen. If the metal does not contain structural cavities near the surface or does not dissolve hydrogen, reaction V also slows down and so does corrosion. However, when the metal dissolves hydrogen, the hydrogen atoms formed at the surface, diffuse inward. If cavities are present near the surface, molecular hydrogen will form there, creating a large pressure in the cavity. This pressure may raise the surface of the metal, a phenomenon known as "blistering". If the cavities are not near the surface, but further away inside the metal, the build-up of pressure in these will not show blistering but will weaken the metal, so that when under any load, it will crack by forces very much smaller than would be needed to break the original metal. These phenomena of blistering and "hydrogen embrittlement" of steel cause considerable anxiety in industry.

Inhibition of reactions IX or X is caused by adsorption of substances such as H_2S or H_2Se on those sites on the metal surface where combination of atomic hydrogen takes place. These phenomena are particularly pronounced in the so-called "sour oil fields" where the water accompanying oil contains H_2S.

2. Corrosion in the Presence of Air

The best illustration of corrosion in the presence of air is obtained by placing a drop of electrolyte solution on a clean surface of iron. After several days one would see that the centre of the circle, created by the solution, is corroded away, while the periphery of that circle is bright; between the two areas would be a circle of brown rust (see Fig. 97). The area corroded away is obviously the anode area, the area covered by the solution which remained bright is the cathode where oxygen was being reduced, and the circle of rust

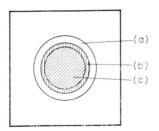

FIG. 97. (a) Cathode area; (b) Rust formation; (c) Corroded anodic area.

is the area of contact between the cathode and anode areas. The reactions of this cell are

$$\text{anode} \qquad Fe = Fe^{2+} + 2e \qquad\qquad \text{XII}$$

$$\text{cathode} \qquad O_2 + 2H^+ + 2e = H_2O_2 \qquad\qquad \text{XIII}$$

$$Fe^{2+} + H_2O_2 = Fe(III) \text{ oxides (rust)} \qquad\qquad \text{XIV}$$

The periphery of the drop is more accessible to oxygen diffusing from outside and therefore acts as the cathode; the centre receives no supply of oxygen, reacts as the anode and so corrodes. Obviously, for best results one should remove the oxygen from the electrolyte solution before putting the solution on the iron surface. If oxygen saturated electrolyte were used we would initially see that the areas of the anode and cathode were not very well defined until all the oxygen has been consumed. Then the electrode areas would be better defined and the picture, as illustrated in Fig. 97 will emerge.

This simple experiment illustrates many points of interest of corrosion in air. It shows that even very pure materials are prone to corrosion if conditions are such that one part is more accessible to oxygen than the other. Thus, if a good supply of oxygen is kept over the whole area of the metal corrosion would be prevented. If, however, the oxygen supply is good except to some small area, such as when an object is placed on the metal surface (Fig. 98), corrosion will be limited to those oxygen-poor areas, but since

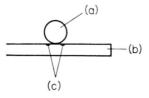

FIG. 98. Corrosion in oxygen poor areas; (a) Plastic bar; (b) Metal plate; (c) Corrosion.

the area of the anode is very much smaller than the area of the cathode, corrosion will be very deep, resulting in rapid deterioration of the metal in the contact area.

Another type of corrosion in the presence of air is direct oxidation. Most metals are less stable than their oxides under normal conditions and, therefore, oxidize when exposed to air. The fact that metals can and have been used extensively without obvious oxidation is due to the formation of a surface film, which insulates the metal from oxygen, stopping corrosion. Let us now see when films form and under what conditions they prevent corrosion. When a bare metal surface is exposed to oxygen, the latter is adsorbed on the surface and reacts there to produce solid metal oxide. If the volume of the oxide is larger than that of the metal, the film formed in this way will cover the metal surface completely, protecting it from further oxidation. In such films, the rate of film growth depends on the rate of diffusion of either oxygen inwards or metal outwards. The rate of diffusion, i.e. the rate of increase of film thickness, is inversely proportional to the thickness of the film, δ

$$d\delta/dt = x/\delta \qquad\qquad 6.9$$

where x is a constant. Integrating gives

$$\delta^2 = 2xt + x_0 \qquad\qquad 6.10$$

where x_0 is the integration constant. Equation 6.10 is that of a parabola, and is called the "parabolic law of growth". Most metals under normal conditions of temperature (i.e. not far from room temperature) and pressure obey this law, but in some cases, the formation of the film does not slow the rate of oxidation and a linear rate of growth is observed. When the film is subject to heating and cooling cycles or to stress, it may break, exposing the metal to further attack and the rate of growth increases, then is slowed down again as the break is repaired. The resulting growth curve looks like a few parabolas joined together (Fig. 99).

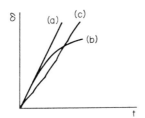

FIG. 99. Growth patterns of metal oxide films; (a) Linear growth; (b) Parabolic growth; (c) Interrupted growth.

Films (of oxide, sulphide or other compounds) may grow on the metal surface either by diffusion of the oxidant inwards or by diffusion of metal ions outwards via crystal vacancies. Both mechanisms operate simultaneously. The resistance of alloys to oxidation depends on the composition in several ways, one of which is known as Hauffe's principle. This principle applies only to metals whose oxide dissolves the cations of the alloying element and where the oxide is oxygen rich, i.e. has many cation vacancies in its structure. Although of limited application, this principle is interesting in revealing the mechanism of metal diffusion. When dealing with, say, nickel surface, NiO is formed which is deficient in nickel, i.e. it contains many cation vacancies. When this oxide is alloyed with lithium, there are now two lithium ions for every nickel ion, therefore there are fewer vacancies. As a result the rate of oxidation of nickel is lowered when a small amount of lithium is alloyed in it. On the other hand when the nickel alloy contains molybdenum, there are more cation vacancies and the rate of oxidation is correspondingly increased.

When a metal is exposed to oxygen containing solution (electrolyte or pure water) so that the whole area is uniformly accessible, without the conditions for the formation of differential areation cells, the whole area may be covered with a corrosion resistant film of oxide, thus retarding the rate of the anodic reaction. This situation can also be obtained by treating the metal with oxidizing agents, such as concentrated nitric acid. The phenomenon, called "passivity", will be discussed in greater detail after the description of corrosion by electric current.

3. Corrosion by Electric Current

Whenever two dissimilar metals are in contact with each other and in a reasonable conducting solution, a galvanic cell is set up; the nobler metal acts as a cathode and the base metal reacts as an anode, undergoing rapid corrosion. The corrosion of the base metal is much slower when it is not in contact with the other metal. The cathodic metal, on the other hand, corrodes only very slightly because it is protected. Good examples of this phenomenon are given by copper plated iron and by zinc plated iron. Copper forms a very good protective plating on iron, as long as there is no break in the plate. If the coating breaks, corrosion of the iron quickly starts and the place of the break soon becomes evident by the rust formed near it. When iron is plated with zinc, there is physical protection by the coating, but when the plating breaks, it is the zinc that undergoes corrosion and not the iron. Hence, the iron article is unaffected until a large portion of the zinc coating has dissolved. The reason for this behaviour is that two metals in contact will strive to achieve equilibrium, but this can only be attained when the activities of the reduced and oxidized species at the electrodes (in the example considered these are the metal ions and reduction products of oxygen) correspond, by the Nernst equation, to the same potential on the cathode and the anode. This equilibrium potential will be E_p, the potential of corrosion. The rate of corrosion, or the current passed, is determined by the current potential curves (not by the current-density potential curves, electrode areas must be considered) which, in turn, are determined by the nature of the electrode process and the state of the electrode surface. Thus, iron will always corrode when in contact with copper, but when in contact with tin it may act either as the anode or the cathode, depending on the circumstances.

Corrosion can also be promoted by the change of environment of the same metal. Consider, for example, the general cell

$$M|\text{aqueous } MX|\text{non-aqueous } MX|M \qquad \text{XV}$$

The difference in activity coefficient between the aqueous and the non-aqueous media is enough to confer some potential on the electrodes and cause corrosion of one of them. This sort of effect is observed in metal pipes that run a considerable distance and pass from one kind of soil into another. The resulting corrosion is then due to what is called "long line currents". When it is possible to put the pipes into the same type of soil all the way, e.g. by filling the trenches with sand or gravel before laying the pipes, this corrosion is prevented.

Another type of corrosion prevalent in buried pipes used to be very frequent in towns where the public transport service was by electric trams. The trams run by taking the electricity from the cable overhead and transferring it to the rails. A portion of the current would leak from the rails into pipes buried in the vicinity, causing intense corrosion of those pipes. This type of corrosion, because of stray current, should be looked for in any environment where large metal bodies are present not too far from sources of large currents.

4. Passivity

A metal is called "passive" when it does not liberate hydrogen when placed in dilute acid solutions and does not corrode in environments that would usually be corrosive. Very many metals exhibit passivity if the correct conditions of pH and potential are selected. For example, chromium is passive over a large range of potential and pH, a fact which we utilize when using this metal for protection of iron, by alloying or plating. Iron, on the other hand, is passive only at high potential in moderately alkaline solutions.

The relationship between the potential, pH and the species of greatest thermodynamic stability can be plotted in a "Pourbaix diagram", a general form of which is shown in Fig. 100. The section marked "immunity" corresponds to the stability of the metal at that region of potential and pH. This is the region of negative potential, often independent of pH in the acid region but dependent on it in the alkaline region. As the potential becomes more positive, i.e. we move up the diagram, compounds of the metal become more stable. In acid solutions, i.e. at the left hand side of the diagram, the most stable compound is usually the hydrated ion (Pourbaix diagrams, it should be remembered, are plotted for an aqueous solution containing no complex forming ligands). As we move right on the diagram, toward the neutral pH, the oxide or hydroxide of the metal becomes the most stable compound. Often this oxide, being insoluble, forms a protective film on the metal, diminishing its chemical activity. This is the passive region. At very

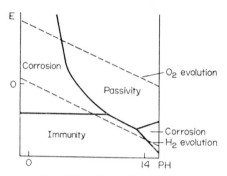

FIG. 100. A Pourbaix diagram.

high pH, many metals form oxy anions, the best known of which are zincate and alluminate, but iron also forms ferroate and ferrite ions in concentrated alkali at fairly low potential and ferrates at high potentials. Thus, the extreme right side of the diagram is often, but not always, a corrosion region. Two oblique lines are drawn across the diagram to indicate the region of the thermodynamic stability of water. From Fig. 100 we see that the region of immunity corresponds to potentials more negative than that of hydrogen evolution. If corrosion processes were determined only by thermodynamic this would have meant that if we apply the immunity potential to the metal, we would liberate hydrogen on it. Fortunately, hydrogen has a considerable overvoltage on many metals, so that the immunity potential is often accessible.

Pourbaix diagrams give us the regions where the various compounds are thermodynamically stable. This is very helpful information, but is often not

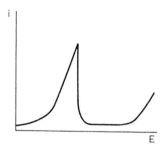

FIG. 101. Current potential curve showing passivity.

enough. Information on the kinetics of the various reactions forming those compounds is also necessary. This is given in the form of current potential curves, a typical curve is shown in Fig. 101. We see that at low or negative potentials there is no current, corresponding to the region of immunity. As the potential becomes more positive, the current increases, showing the formation of a soluble corrosion product. On increasing the potential the current drops to very low values. The maximum value of the current before it starts to fall is the passivation current. The region where the potential is highly positive but the current is very small is the passive region. The current increase which follows the passive region may be due to the oxidation of water (evolution of oxygen) but sometimes it is due to the formation of a high valence state of the metal, e.g. chromium is oxidized to chromate. Clearly, pH is not considered in such curves, a separate current potential curve has to be taken for each pH.

5. *Protection against corrosion*

There are three basic ways by which corrosion is prevented: (*a*) by physical isolation of the metal from the oxidant, (*b*) by reducing the rate of the cathodic reaction using any of the methods known in electrodics and (*c*) by maintaining the metal at a potential in the immunity region of the Pourbaix diagram.

Physical isolation of the metal from the oxidant is attained either by painting it, coating it with another corrosion resistant metal or forming on it a protective, usually oxide, film. Painting a metal is usually preceded by the process called "pickling". This involves washing the metal in acid in order to remove any existing corrosion. Acid, however, is itself a very corrosive medium for most metals, therefore we add to the pickling solution a compound known as a "pickling restrainer". This compound is adsorbed on the active sites of the metal and poisons them, rendering the metal stable in the pickling solution and painting is then done on clean, wholesome surfaces.

When protective films are easily formed on the metal surface by exposure to air, the metal is naturally in the passive state and corrosion resistant. Many metals, however, have to be protected by alloying with one of the naturally passive metals, e.g. chromium or aluminium. In other words, the naturally passive metals have very low passivation current; they form their protective coating rather quickly when current is applied to them. When they are alloyed with corroding metals, they lower the passivation current of the latter and help to form a pore free inhibitive oxide layer.

Passivity can also be achieved by placing the metal in a strong oxidizing agent, such as conc. HNO_3 chromate or permanganate salts solutions. The effect is exactly the same as when a high potential is applied to the metal. It is interesting to note that metals whose oxides are more acidic than basic, such as tantalum, molybdenum and tungsten, have excellent stability because the oxides formed on their surfaces are insoluble in acid (although it is of course soluble in alkali) as are the oxides of many other metals. Metals particularly resistant to alkali are nickel, magnesium and silver. So far we have emphasized the formation of oxide films. There are, of course, many compounds that form insoluble films on metals and can be used as corrosion preventors. As examples one could name lead sulphate or silica which is formed on iron in the presence of sodium silicate.

Another form of protective coating is the plating of one metal with another which is corrosion resistant. Most commonly this plating is just a physical separation of the corroding metal from the oxidant, but sometimes, notably in the case of galvanized iron (zinc plated iron), an electrochemical process is also involved. As a rule metals thus coated are protected against corrosion as long as the coat is intact. When it breaks, corrosion at the fissure is likely to proceed even faster than had the metal been uncoated, because of the creation of a galvanic cell at the fissure. This cell may be of the differential areation type or of the two dissimilar metals in contact type. Clearly, coatings require careful maintenance or replacement by a different method of protection.

The second method of protection is primarily limiting the rate of the electrode reaction, commonly the cathodic one. Obviously, in an oxygen free, neutral pH system, metals will last indefinitely, but these systems are not of practical value. Using metals with a high overpotential towards hydrogen evolution, such as lead, in acid solutions is a good technique in preventing corrosion by acids. Often one uses materials, known as "retardants" which adsorb on the metal and retard the hydrogen evolution or oxygen reduction. The use of retardants to check acid corrosion needs care because of the blistering and embrittlement phenomena discussed above. One way of overcoming these latter phenomena is by compressing the surface by proper metallurgical treatment thus eliminating the accumulation of hydrogen in cavities near the surface.

One way of reducing the rate of corrosion is by reducing the area of one of the electrodes, commonly the anode in a potential differential areation cell, or by eliminating it altogether. Thus, the anodic reaction is checked by ensuring a uniform accessibility of air to the metal surface.

The third and final way by which corrosion is prevented is by "cathodic protection". We can bury a conductor in the ground, next to the system of pipes to be protected or to the other installation, and connect a d.c. power supply across the pipes and the conductor, in order to make the pipes cathodic. In other words, the potential applied to the pipes is in the region of immunity in the Pourbaix diagrams. This results in complete protection of the pipeline. However, caution must be exercised here. Figure 102 illustrates the pipeline to be protected, the buried conductor and another system of pipelines, between these two. This second system will now be

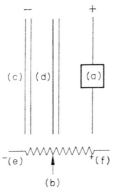

FIG. 102. Inadequate cathodic protection of pipes; (a) Power supply; (c) Protected pipe; (d) Corroding pipe; point (b) is positive with respect to point (e).

anodic to the protected one, resulting in considerable damage to this second system. This problem can easily be solved by connecting all the pipes to the same protecting system. This, however, is an administrative problem rather than a chemical one.

Sometimes it is not practical to build a separate conductor and power supply because the system to be protected is too small. In this case we use sacrificial anodes. The system (e.g. a steel water heater) is connected to a metal rod, such as magnesium or zinc, which will corrode in preference to steel. Thus the protected installation is the cathode of the created cell and does not corrode. The sacrificial anodes must, of course, be checked and replaced periodically.

D. Batteries and fuel cells

In 1800 Volta described the first "voltaic pile" or "galvanic cell", i.e. a source of direct electric current at low potential. The first cell was made from discs of copper and zinc separated from each other by a cloth saturated with salt solution. Ever since then, the search for better cells to supply electricity has continued. This section will start by classifying the various kinds of batteries and fuel cells, proceed to describe some of the principles of operation and define the efficiency of the cells, and then describe the general features of each kind of cell.

The word "battery" originally meant several galvanic cells connected together to provide more energy than is obtainable from a single cell. Today this word means either a single galvanic cell or several of them. "Fuel cell", on the other hand, means just one cell; when several are connected and function together they are called a "fuel cell battery" or a "fuel battery".

The ideal cell should give high currents for any given voltage, should be inexpensive to manufacture and operate, and should not be too bulky or too heavy for the application at hand. The galvanic cell is the reverse of the electrolytic cell—the reduction at the cathode and oxidation at the anode occur spontaneously. The reactants can be part of the construction of the cell or supplied from the outside. The reactants in batteries are part of the construction, they change as the cell reaction takes place and products accumulate; the life of batteries is limited. Fuel cells operate on the principle of unchanging electrodes and unchanging electrolyte, the reactants are pumped into the cell and the products are lifted out. Thus, the life of a fuel cell is theoretically infinite. In practice, the accumulation of impurities and corrosion products does limit the useful life of the fuel cell.

All galvanic cells are divided (as has by now become apparent) into batteries and fuel cells. Batteries are themselves divided into primary and secondary. In primary batteries, the cell reaction is irreversible, in secondary batteries the reactants can be regenerated by passing direct current to reverse the polarity of the electrodes†

Fuel cells have been classified according to many criteria. Here we shall adopt the classification according to temperature of operation: low temperature cells operate close to ambient temperature; medium temperature cells operate between 100°C and 600°C and high temperature cells operate close to 1000°C. Low temperature cells use mainly aqueous electrolytes, medium ones use either very concentrated acid solutions or molten carbonates and high temperature cells use solid oxides as electrolytes. The fuels are usually gaseous, sometimes liquid; solid fuels are unimportant in present day fuel cells.

1. Principles of Operation

The reaction in any galvanic cell can be generally presented as

$$A + B = \text{products} \qquad\qquad \text{XVI}$$

and a certain change of enthalpy ΔH and of free energy ΔG is associated with it. The enthalpy is related to the heat produced by the reaction, e.g. the

† Since the regeneration is carried out in the cell, it must periodically be taken out of service. To make the cell useful for a longer part of the time, some experimental schemes have been developed for regenerative fuel cells: the cell reaction products are removed, regenerated and recycled.

calorific value of a fuel is given by the enthalpy change associated with its combustion reaction. The potential of the cell is related not to the enthalpy, but to the free energy change. When comparing galvanic cells to steam turbines or internal combustion engines for efficiency and performance this difference must be borne in mind. The value of the entropy of the cell reaction is thus important in evaluating the efficiency of the cell.

$$\Delta G = \Delta H - T\Delta S \qquad 6.11$$

When the value of ΔS is positive, the potential of the cell is equivalent to less energy units than are obtained by direct combustion as heat. We can define a factor to compare the amounts of energy produced by these two processes. The "comparative thermal efficiency" of a cell reaction is

$$\varepsilon_r = \Delta G/\Delta H \qquad 6.12$$

If ΔS is positive ε_r is less than one (or less than 100%); if ΔS is negative ε_r is more than one. ε_r is a theoretical factor and as such is not related to the design of cells. In operating cells it is found that the experimental voltage may be less than the one calculated from the Nernst equation. This is usually due to the existence of a mixed potential at one or both electrodes. A mixed potential arises when there are several processes taking place simultaneously at an electrode although the nature of all mixed potentials is not usually known and, therefore, cannot be calculated. Therefore, it is useful to define the "voltage efficiency" of a cell

$$\varepsilon_v = V'/V \qquad 6.13$$

where V' is the experimental voltage at zero current and V is the theoretical one.

The current passed through the cell may be due to several electrochemical reactions. If one of these reactions involves fewer electrons than expected, then the current observed will be less than expected and the current efficiency will be

$$\varepsilon I = I'/I \qquad 6.14$$

where I is the expected current and I' is the observed one. An example of low current efficiency is the oxidation of methanol. Normally methanol is expected to be oxidized to carbon dioxide and water

$$2CH_3OH + 3O_2 = 2CO_2 + 4H_2O \qquad \text{XVII}$$

involving six electrons per methanol molecule. One of the side reactions possible at a methanol anode is the formation of formate or formic acid

$$CH_3OH + O_2 = HCOOH + H_2O \qquad \text{XVIII}$$

involving only four electrons per methanol molecule. In practice, the amount of formic acid formed in acid electrolytes is small, but in alkaline solutions most of the product is formate ions. Under such conditions the current efficiency is 66·7%.

The efficiency of galvanic cells, the permissible current and operating voltage, as well as the power of the cell, can all be calculated from current potential curves of the two electrode reactions. Figure 103 shows typical

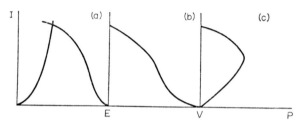

FIG. 103. Characteristic curves of a fuel cell as a function of current; (a) Electrode potentials; (b) Cell voltage; (c) Cell power.

current potential curves for a cathode and an anode. At zero current the voltage of the cell is maximum, as indicated. The potential decreases as the current increases, as shown in Fig. 103(b), but there always exists a definite voltage across the cell which is determined by the current drawn and the current potential curves of the two electrodes. However, when one of the electrodes reaches the limiting current region, it no longer has a unique value of potential. Correspondingly, the cell using such an electrode would not have a unique voltage, it will "collapse" and the voltage will reduce to zero. Obviously, this situation should be avoided and therefore we provide for as large a current as possible by using high concentrations of reactants and large electrode areas.

The power of the cell, i.e. $I·V$, is shown in Fig. 103(c) as a function of cell current. We see that this function passes through a maximum which is given by the current-voltage curve of the cell. This point of maximum power can be regarded as the best operation point, provided no other considerations dictate other values for cell voltage or current as the most suitable.

From Fig. 103(a) and (c), we see that the power of the cell is high if the current potential curves of the separate electrodes are steep. A steep current potential curve implies a small overpotential, i.e. a fast electrode process. If the electrode process is not naturally fast, its rate can be increased by one of two ways: a catalyst may be added to the electrode, or the temperature of operation may be raised. The catalyst used is chosen according to the reaction to be catalysed, i.e. a different one will be used for the decomposition of peroxide than for the reduction of metal oxide. Therefore, the choice of a suitable catalyst may be either empirical or dictated by the nature of the rate determining step in the electrode reaction. The merit of medium and high temperature cells is that the electrode reactions proceed rapidly at those temperatures, making catalysts (which are usually expensive) unnecessary.

The resistance of the electrolyte also reduces the current obtained from galvanic cells. This consideration is of prime importance in the design of cells. The cathode and anode are placed as close to each other as is possible and the electrolyte solution is made as conducting as possible using fairly concentrated solutions of acids or bases in water. In high temperature cells, which usually use solid oxides as electrolytes, IR drop in the electrolyte is the most important factor in determining the power.

As mentioned before, factors other than electrochemical must be considered in the production of fuel cells and batteries. For example, the cost of corrosion resistant casing for the cell, as well as the cost of electrodes, fuel and electrolyte, must be considered. For space applications, cost is not as important as reliability and long life; for the motor industry the prime considerations are power per unit weight and unit volume, as well as cost.

2. *Primary batteries*

The number of different commercial primary batteries is not very large, but their importance, as measured by the number being manufactured, has risen very considerably in the last years, particularly since transistor radios and other portable electronic equipment have become popular.

The most popular battery is based on the Leclanche cell, invented in 1865. The original was a wet cell, containing a solution of ammonium chloride. The modern battery is dry, i.e. the electrolyte is contained in an absorbing medium so that it does not flow. The anode is zinc, the cathode is manganese dioxide and the cell reaction is

$$Zn + 2MnO_2 + 2NH_4Cl = Zn(NH_3)_2Cl_2 + 2MnO(OH) \qquad XIX$$

The performance of the cell is not very good, but its construction is cheap and this accounts for its popularity.

When the ammonium chloride electrolyte is replaced by a potassium hydroxide solution, the result is what is known as an alkaline manganese battery. The electrolyte solution (which is, of course, held in an absorbing medium) contains enough zincate ions to suppress open circuit dissolution of the zinc. The cell reaction is

$$Zn + 2MnO_2 + 2H_2O = Zn(OH)_2 + 2MnO(OH) \qquad XX$$

The performance of this cell is very much better than that of the Leclanche cell, but the price is higher.

The power density of the two batteries described is not enough to cope with miniature circuits, such as in watches and hearing aids. Batteries with very stable voltage and very long life are based on the zinc anode and a mercuric oxide cathode, with an alkaline electrolyte. They are known as mercury cells. The cell reaction is

$$Zn + HgO + H_2O = Zn(OH)_2 + Hg \qquad\qquad XXI$$

It should be noted that in the mercury cell, only one mole of mercuric oxide is needed to produce the same amount of electricity produced by two moles of manganese dioxide in the previous cells. This means that the mercury cell can be made considerably lighter in weight. However, since mercury is expensive, the cell based on it is expensive too and therefore has not become generally popular.

The performance of the mercury cell is very good since it keeps a steady voltage of approximately 1·2 volts over a wide range of current and for a fairly long time.

Other cells, based on zinc anodes or on mercuric oxide cathodes are known. Among them are the silver-zinc battery, zinc-copper oxide battery, mercury-cadmium battery etc.

3. Secondary batteries

This section is concerned with cells where the reactants can be regenerated by one of several means: electrolysis, heat, radiation or chemical reaction. We shall describe the well known storage batteries and go on to discuss other regenerative systems that have not yet become commercially available.

The lead acid battery is the best known and most popular of all secondary batteries, or "accumulators". The cathode is lead metal, the electrolyte is sulphuric acid and the anode is lead dioxide. As current is drawn from the battery, lead sulphate forms on both electrodes.

$$Pb + SO_4^{2-} = PbSO_4 + 2e \qquad\qquad XXII$$

$$PbO_2 + SO_4^{2-} + 4H^+ + 2e = PbSO_4 + 2H_2O \qquad\qquad XXIII$$

Clearly, as the reaction proceeds, the amount of sulphuric acid in the electrolyte is decreased, and the amount of water is increased.

The lead acid battery is used in all cars and in most other places where a portable source of direct current is needed. When the cell reaction has gone almost to completion (but not quite), the terminals are connected to a current source so as to force reactions XXII and XXIII in the opposite directions. Charging of the battery takes several hours, during which time the battery is out of operation. Another drawback of the lead acid battery is that it is a wet cell, therefore, it cannot be handled as easily as one would wish.

Another storage battery which has been developed commercially is the nickel-cadmium battery. In this cell the anode is made of cadmium and the cathode is of nickel hydroxide, which is probably in the trivalent state. The cell reaction is

$$2Ni(OH)_3 + Cd = 2Ni(OH)_2 + Cd(OH)_2 \qquad XXIV$$

The electrolyte is alkaline, and the cell is used as a wet cell. Its performance in terms of long life and power is somewhat superior to that of the lead acid battery, but its price is much higher. The chief merit of this cell is the use of an alkaline rather than acid electrolyte, reducing corrosion, but this is also a problem, since the alkaline solution absorbs carbon dioxide from the air. Carbonates in the electrolyte solution diminish the life of the cell. The cell cannot normally be hermetically closed, because in the last stages of charging hydrogen and oxygen are evolved at the electrodes and these must not be allowed to accumulate. The best way to avoid high pressure in the cell is to keep it open to the atmosphere.

The importance of the nickel cadmium cells lies in the fact that, by a special design, they can be made "dry" giving the portability and freedom of dry primary batteries and the economy of a storage battery. Toward the end of the charging process, both hydrogen and oxygen are liberated at the electrodes, the electrode reactions being

$$Cd(OH)_2 + 2e = Cd + 2OH^- \qquad XXV$$

$$2H_2O + 2e = H_2 + 2OH^- \qquad XXVI$$

at the cathode and

$$Ni(OH)_2 + OH^- = Ni(OH)_3 + e \qquad XXVII$$

$$4OH^- = O_2 + 2H_2O + 4e \qquad XXVIII$$

at the anode. The evolution of hydrogen can be delayed by adding excess $Cd(OH)_2$ to the electrode and the evolution of oxygen can be delayed by adding excess $Ni(OH)_2$ to the other electrode. It is, of course more advantageous to curb the evolution of hydrogen because the quantity evolved is double that of oxygen. Moreover, if the oxygen is allowed to reach the cadmium electrode, it can oxidize the latter directly, thus relieving any pressure in the cell. Indeed, the cadmium nickel cell can be hermetically sealed if excess cadmium hydroxide is present and if there is free access of oxygen to the cadmium electrode. Hermetically sealed rechargeable batteries are now commercially available.

Any rechargeable battery must be taken out of use when it is being recharged and the larger the battery, the longer it takes to recharge it. Thus,

14

a rechargeable fuel cell, where the reactants are reconstituted away from the cell proper, is very attractive. These cells, however, have not yet become commercially available, their cost is still much too high to compete with other power sources. The best known rechargeable fuel cell is the lithium hydrogen cell. This consists of a lithium anode and a hydrogen cathode, the electrolyte is the eutectic mixture of lithium fluoride and lithium chloride at 600°C and the reaction product is lithium hydride. This is pumped to a regenerator, which operates at 900°C, where lithium hydride is dissociated to its elements. After separation the molten lithium and the hydrogen are recycled. Except for the obvious engineering problems involved in the design of a cell at these temperatures and the corrosion problems, there is the very low efficiency of this cell, which has been only 6–10%. Research into the design of regenerative fuel cells is progressing and hopefully that they will become commercial in the not too distant future.

4. Low Temperature Fuel Cells

All low temperature fuel cells use water as the solvent, and acid or base as the electrolyte. The cathode is almost always an oxygen or air electrode and the anodes may use either gas (usually hydrogen) or water soluble fuel, such as methanol, hydrazine or ammonia. This section will describe first the oxygen electrode, then the hydrogen electrode and the hydrogen-oxygen fuel cell. Finally, the methanol consuming anode will be described and some of the problems outlined.

The oxygen gas is reduced at the cathode of the fuel cell via one of two mechanisms: the first is

$$O_2 = O_2(ads) \qquad \qquad \text{XXIX}$$

$$O_2(ads) + 2e = O_2{}^{2-}(ads) \qquad \qquad \text{XXX}$$

$$O_2{}^{2-}(ads) + H_2O = HO_2{}^- + OH^- \qquad \qquad \text{XXXI}$$

peroxy anion is then reduced further to hydroxyl. The second mechanism is

$$2M + O_2 + 2H_2O = 2M(OH)_2 \qquad \qquad \text{XXXII}$$

$$M(OH)_2 + 2e = M + 2OH^- \qquad \qquad \text{XXXIII}$$

The exact mechanism depends on the electrolyte, temperature and electrode material, e.g. on carbon electrode in alkaline medium the first mechanism is the predominant one; on platinum electrodes the second mechanism is the correct one.

It is clear that the potential of the oxygen electrode at open circuit is not the Nernst potential for the straightforward reaction

$$O_2 + 2H_2O + 4e = 4OH^- \qquad \qquad \text{XXXIV}$$

although, with careful experimental technique, the latter can be attained. The catalyst used for the oxygen electrode must either catalyse the decomposition of peroxide or the reduction of metal hydroxide. Since it is not known *a priori* which mechanism is operating under the exact experimental conditions at hand, the search for a suitable catalyst must be empirical, rather than theoretical. In alkaline solutions, it was found that certain spinels, as well as the high oxidation state of nickel oxide can be used as catalyst. In acid solutions the best catalysts are the precious metals—platinum and alloys with gold and irridium.

The concentration of oxygen in the oxidizing gas does not change the reversible potential of the oxygen electrode very much. The Nernst equation predicts that the change in potential for changing from pure oxygen at atmospheric pressure to air at the same pressure is only several millivolts. The change in the rate of the reaction, however, is much more pronounced; while the voltage is given by a logarithmic function of concentration, and therefore, is not very sensitive to changes of the latter, the rate of reduction, and hence the current, is directly proportional to concentration. Therefore, the power of an air operated cathode will be considerably smaller than that of one operated on pure oxygen. The advantages gained by using pure oxygen, however, must be weighed against the cost of obtaining it.

The structure of the electrode used for reacting gases in general, and of course oxygen as well, is another problem. This electrode has to maintain as large an area as possible of the interphase gas–electrolyte–electrode. Therefore it is usually made of a porous material where the liquid phase occupies a portion of the pores and the gas phase occupies the other portion. For a material of given porosity the area of contact (Fig. 104) becomes larger as the size of the pores becomes smaller. In very small pores there exists the danger of flooding the electrode with liquid by the capillary action of the pores if special precautions are not taken. This is an ever present problem in the construction of gas electrodes and is sometimes solved by treating the electrode with a water repellent material, or by a special construction. One

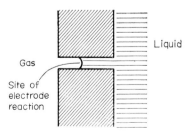

FIG. 104. Site of a gas electrode reaction at the gas-liquid-solid interphase.

of the most popular such constructions is the biporous electrode. This is composed of two sheets of porous metal—one has the larger pores and contains the gas, the other is of very small pores and contains the electrolyte. The difference in the capillary forces of the two sheets hold the phases apart and in contact. This construction is also used at high temperatures, where water repellent agents are no longer useful.

The hydrogen electrode poses many theoretical problems. The exact mechanism and the identification of the rate determining step are still unknown in spite of the effort made in studying them. However, from a practical point of view, this electrode is easy to make since all the platinum metals have a very high exchange current density for hydrogen evolution. Obtaining cheaper metals which can work as well is a more difficult problem. The electrolyte can be either acidic or alkaline, although the performance of the electrode seems to be better in acid solutions.

The hydrogen oxygen fuel cell is the most developed and, so far, the only one used as a source of power outside the research laboratory. This cell, when made into a fuel battery by connecting it to other cells, together with the high pressure gas cylinders needed, is much lighter than the conventional lead acid battery, thus making it a more desirable power source where electric power is needed but lightness is of primary importance.

As is clear from the previous discussion, the use of gas as a fuel presents problems of electrode construction and storage, that are not there when the fuel is water soluble. Three substances have been seriously suggested as such fuels—methanol, hydrazine and ammonia. Of these, methanol will probably become more accepted than the other two, mainly for economic and safety reasons. Methanol, if it becomes widely used for fuel cell operation, is likely to drop in price markedly, while hydrazine and ammonia will probably not. This section will describe briefly the methanol fuel cell, although the technology of the hydrazine one is more advanced.

The physical properties of methanol, its melting point, boiling point and solubility in water, approach the ideal for soluble fuels; it is also resistant to reduction at the cathode. However, it is inflammable and, although it can be electrochemically oxidized to carbon dioxide and water, its exchange current density is small and suitable inexpensive catalysts to yield high current densities are not yet known. This latter drawback is the one holding back the commercial development of the methanol fuel cells.

The electrode reaction of methanol in acid solutions is

$$CH_3OH + H_2O = CO_2 + 6H^+ + 6e \qquad \qquad XXXV$$

In alkaline solution, the carbon dioxide does not escape as gas, but forms carbonate ion

$$CH_3OH + 8OH^- = CO_3^{2-} + 6H_2O + 6e \qquad \qquad XXXVI$$

The current potential curves of methanol electrodes show the expected rise in current when the potential is increased, followed by the limiting current region. This limiting current was shown to be diffusion controlled if the concentration of methanol is under 0.8 moles litre^{-1}. Above that, the limiting current is no longer proportional to concentration, but increases at a much lower rate than the concentration. This shows that the electrode reaction might be limited by a preceding chemical reaction. When the potential of the electrode is held constant, the current drawn declines rapidly in the first few moments, then more slowly over a few hours. This phenomenon is also observed in fuel cells, where the performance of the cell deteriorates with time. This situation can be corrected by interrupting the current; after reconnecting the leads, the cell performs well again. Another troublesome phenomenon connected with methanol fuel cells is that of electrical oscillation. At a certain point on the current potential curve, steady values of current (at controlled potential) or potential (at controlled current) cannot be obtained, the current fluctuates with time.

The oxidation product of methanol on electrodes in acid solutions was shown to be carbon dioxide and water. However, in alkaline solutions, the formate ion formed resists further oxidation, and sometimes the sole product is formate. This reduces the current efficiency and causes problems of changing the electrolyte when the concentration of formate is too high.

The several catalysts which have been used for the electro-oxidation of methanol are the platinum metals and nickel. It seems that platinum, again, is the best catalyst known, here as in the hydrogen oxygen cell. The mechanism of the electrode reaction is complex and involves several consecutive reactions. It also depends on the exact conditions of the experiment.

5. *Medium Temperature Fuel Cells*

When natural gas is considered as a possible fuel, an increase in the rate of the electrode reactions is needed. Since catalysts are either very expensive or unknown, temperature is raised to lower the overpotential. Since the products of oxidation of natural gas are carbon dioxide and water, these will always be present in the gas mixture over the cell. An equilibrium between the gases and the molten salt electrolyte will be established, and part of the electrolyte will be converted to carbonate, regardless of the nature of the original anion. Therefore, it seems reasonable to use molten carbonates, not other salts, as electrolytes. On the other hand, since carbonates dissociate at high temperatures to give carbon dioxide, it is necessary to keep the partial pressure of CO_2 above the cell at such a value as to retard any change in the composition of the electrolyte. Both the fuel gas and the air are premixed with carbon dioxide before being fed to the fuel cell.

Molten carbonate cells have been operated on a variety of fuels: carbon monoxide, hydrogen, kerosene and a variety of hydrocarbon gases mixed with steam. The reaction with carbon monoxide is

$$O_2 + 2CO_2 + 4e = 2CO_3^{2-} \qquad \text{XXXVII}$$

at the cathode and

$$2CO + 2CO_3^{2-} = 4CO_2 + 4e \qquad \text{XXXVIII}$$

at the anode. The transfer of oxygen from air to carbon dioxide proceeds via the carbonate ions. The reactions with the other fuels are, of course, different, but in all cases the final product should be carbon dioxide and water.

The fused carbonate cells operate between 500 and 800°C. In these temperatures the main feature that limits the current density at a given potential is not activation polarization or diffusion current, but the internal resistance of the electrolyte. Another persistent problem is the extreme corrosive nature of the electrolyte, limiting the choice of electrodes to those which resist corrosion, i.e. the noble metal ones.

There are basically three types of construction of molten carbonate cells. In the most developed type, the electrolyte is contained in a porous diaphragm of magnesia. This type of construction reduces corrosion, but increases electrolyte resistance. The second type, that of free flowing electrolyte, has not been developed because of the serious corrosion problem at the temperatures of operation of these cells. The third type uses electrolyte which is mixed with magnesia powder to a stiff paste. This structure seems to have the merits of both the other structures in that it reduces corrosion without much affecting the resistance of the cell.

The main factor that limits the life of molten carbonate cells is that of deterioration of the construction materials. Unless these cells can be more cheaply produced or of much longer life, it seems that they will be abandoned in favour of the solid oxide cells which will be described in the next section. Another disadvantage of the molten carbonate cells is the high temperature of operation. Experimental cells using concentrated aqueous acid solutions and operated at less than 200°C will be described presently.

It has been observed that light hydrocarbons, such as ethane propane etc., can be oxidized on platinum black in strong acid electrolytes. The original electrolyte studied, sulphuric acid, was suitable only for cells operating under 100°C. At higher temperatures direct reaction took place between the acid and the hydrocarbon. Phosphoric acid, however, not being an oxidizing acid is suitable for cells in the temperature region of 150–200°C. Using such hydrocarbons in phosphoric acid electrolyte gave electrodes capable of producing rather large current densities at not too large polarizations. However, there are two very serious drawbacks to such an electrode. The first is the nature of the catalyst (platinum black was found to be the best) and the second is the very corrosive nature of phosphoric acid at these temperatures. Thus, molten carbonate cells and strong phosphoric acid cells seem to be in the same category and their future depends very much on research now in progress.

6. High Temperature Cells

With high temperature fuel cells, we consider the use of solid oxide electrolytes at 1000°C. The solid oxide needs to be made conducting, must be stable toward oxidation and reduction, must be physically stable, i.e. should not crack or disintegrate on repeated heating and cooling, and the conductance of the oxide must be anionic, i.e. it should be entirely by the movement of oxide ions since any cationic mobility would result in permanent change in the composition of the electrolyte. Oxide ion movement does not result in any change in the electrolyte because of the use of oxygen electrodes; oxide ions are generated at the cathode and combine at the anode to form carbon dioxide. The best solid oxide electrolyte found so far is zirconia (ZrO_2). In the untreated state it is very unstable, and suffers considerable change in dimensions on repeated heating and cooling. It is also a typical insulator, its specific resistance at 1000°C is approximately 107 ohm cm. All these undesirable properties are eliminated by introducing other oxides, calcia and yttria, which have a cubic structure like zirconia and form solid solutions with it. This "stabilized" zirconia is very inert and mechanically stable; it is also a good conductor, having a very small resistance at high temperatures. The high conductivity is probably because the stabilized zirconia, when assuming the structure of the pure oxide, must leave some vacant oxide sites in order to preserve the electrical neutrality. These sites are responsible for the conductance by migration of oxide ions.

The fuel in these high temperatures cells is a mixture of gaseous hydrocarbons, but these are not stable at the temperatures of operation. Cracking reactions occur readily, resulting in a mixture of low hydrocarbons and hydrogen and deposition of carbon in the pores of the electrode. This deposited carbon is not reactive and clogs the pores of the electrode, leading to deterioration of the cell. Treatment of the hydrocarbons with steam, however, results in the following reactions

$$C_2H_6 = C_2H_4 + H_2 \qquad \text{XXXVII}$$

$$C_2H_4 + 2H_2O = 2CO + 4H_2 \qquad \text{XXXVIII}$$

$$CO + H_2O = CO_2 + H_2 \qquad \text{XXXIX}$$

which are called, respectively "cracking", "steam reforming" and "water gas shift". The result of these reactions is that the only materials in the electrode pores are gases. The low hydrocarbons, carbon monoxide and hydrogen react in the normal way and carbon dioxide is evolved as waste. Thus "steam reforming", which can be done in a simple chamber packed with nickel gauze or turnings for example, improves the performance of hydrocarbon electrodes. The exact temperature of the steam and the cell, the amount of steam and the composition of the resulting gas mixture can be calculated from phase diagrams of the carbon–hydrogen–oxygen system.

The electrodes in high temperature cells can be made of several metals or carbon and are usually sprayed or vacuum deposited on a thin wafer of the oxide electrolyte. This results in a porous electrode structure, needed for the penetration of the gas, and a good electrical contact between the electrode and the electrolyte. The thickness of the electrolyte must be very small, thicknesses under two millimetres have been successfully made and employed.

High temperature fuel cells, employing a solid electrolyte are much more convenient than cells using molten carbonate or concentrated phosphoric acid electrolytes since they are dry and non-corrosive. The need for steam reforming of hydrocarbon fuel is not unique to the high temperature cells, molten carbonate cells also perform better when the fuel is steam reformed first. The only severe problem of high temperature fuel cells which hinders their wide application is that of heating them to the correct temperature and maintaining it—it is not at all feasible to have to heat one's car to 1000°C every morning before using it.

7. Prospects

Let us divide the discussion in this section to the prospects of fuel cells and those of batteries.

In evaluating the prospects of fuel cells, one must consider their power output and efficiency as a function of their size and compare it to that of conventional power plants, such as internal combustion engines or steam turbine power stations.

The overall efficiency of fuel cells increases with size because problems of heat loss or of gas pumping can be dealt with properly only if the size of the fuel battery is above a certain minimum of a few kilowatts. Above this output there is little increase in efficiency as the size is increased; the efficiency of steam turbine power plants increases with size up to about 500 megawatts. This consideration alone implies that fuel cells will be best used for small scale operations, such as supplying electricity in remote areas or powering vehicles, such as delivery vans or locomotives. They will also be used where the advantages of fuel cells, the quiet operation, reliability and absence of pollution, will be more important than efficiency.

It would be advantageous to evaluate each type of fuel cell separately. The hydrogen oxygen fuel cell is the most highly developed of all and has been used in space exploration as a source of electricity. When power is required for more than a few hours, the weight of this fuel battery with the stored gases is less than that of lead acid accumulators. Another advantage is the speed in which the gas cylinders can be replaced. The hydrogen air cells can be used as a substitute for gasoline operated electricity generators in places where the initial cost of the equipment is not of prime importance. Hydrazine and methanol operated fuel batteries are much more easily engineered than gas operated cells, but the price of catalysts (in the case of methanol) or the fuel itself (in the case of hydrazine) makes the application of such cells feasible only in very special cases.

High or medium temperature hydrocarbon burning fuel cells are the least developed and it seems that they will be suitable for small applications where instant starting is not essential, such as the running of locomotives or plants for the generation of direct current for electrolytic purposes. These cells could also be used in conjunction with nuclear power plants and use the surplus heat furnished by the latter.

While intensive research effort was invested in the fuel cell field, batteries of every size have become very popular. The very small ones are used to power modern electronic portable equipment (and children's toys) and huge lead acid batteries have been introduced to power some delivery vans and other equipment which stop and start much of their working time. At the same time, research on high power density batteries has developed, but has not created the same interest and enthusiasm as the research into fuel cells. While many books have been written on fuel cells, only one has been published on the latest developments in the battery research field.

The main advantage of both batteries and fuel cells is in their small size being very suitable where the electric power needed is small or very small. Small power stations in remote places could be constructed using fuel cells; very small amounts of electric power are conveniently supplied by batteries.

E. Electrochemical production

The term used at head of this section needs definition. Let it include in this context all those commercial processes which use electrical current for the production of materials. The chlor-alkali industry clearly comes under this heading, as does electrowinning and electrorefining of metals. Production of fluorine and fluorinated organic compounds will also be dealt with here. Other uses of electrode processes, such as electrochemical machining and electroforming are not included in this section.

The electrode processes used in industry are fairly simple and straight-forward. The problems involved are mostly of engineering and result when laboratory processes are scaled up. It is instructive to compare laboratory and factory operations with those of kitchens in homes and those in big restaurants. The utensils in a restaurant must be much more robust; many of them are considerably bigger than those at the home. The operations will be different; in the home much of the waste is disposed of in the kitchen sink; in a restaurant this is unthinkable and special processes have to be designed for proper disposal. The method of heating may be different—steam heating is widely used in big establishments, but never in the home.

Returning to chemistry we find the same process—utensils in a factory must be bigger and stronger than those in the laboratory and waste disposal is one of the problems to be dealt with in the plant. In the laboratory, materials are always made in batches, enough can usually be made in one or two because the quantities needed are not very large. In industry it is much more convenient to have flowing systems, such as an assembly line, where one feeds the reagents into the beginning of the line and carries out several operations along it. Naturally, not all the starting material will react, at the end of the line there would be some unreacted starting material as well as product. Separation of the two and recycling of the unreacted portion is another problem of the engineer.

A chemical plant uses starting materials that are not very pure—chlorine is produced from the crude sea salt; metals are produced from ores. Dealing with impurities is another problem for the chemical engineer. Sometimes, as in electrowinning, extensive purification has to be carried out before electrolysis. In other cases, such as electrorefining, electrolysis takes care of impurities and dealing with these is really a problem of waste disposal or recycling of useful material.

Often the most important consideration of the engineer is that of cost, capital cost of buildings and equipment and running cost of replacements and raw materials. Thus, platinum electrodes, although they are the best electrocatalyst, cannot be used in industry because of their price. Corrosion is a prime factor in the running costs of a plant.

Since every plant is constructed for a specific process often using specific raw materials, the design of cells, nature of electrode materials and the processes which accompany the electrolysis step vary considerably. However, there are some features which are common to most electrolytic processes These will be discussed first and later several processes will be briefly described.

1. Cells, Electrodes and Electrolytes

The cell can be considered as a box in which electrolysis takes place. Clearly, this box has to accommodate the cathode and anode, the electrolyte and a suitable device, such as a diaphragm, which separates the two electrode compartments. This separation is essential when the two reaction products react with each other. The cell should also have provisions for the introduction of reactants and removal of products, preferably in a continuous process. This box should be built of materials that withstand the conditions of operation and the current should be supplied via massive bus-bars.

The voltage of an industrial cell, like that of its laboratory counterpart, is composed of the reversible voltage, given by Nernst equation, the electron transfer overvoltage, given by the Butler–Volmer equation, concentration overvoltage, due to mass transport, and ohmic resistance of the electrolyte, electrodes and connections. Clearly, one would try to obtain as much product (i.e. current) for the electrical power invested, as possible. This means that the voltage across the cell should be as low as possible, provided a reasonable rate of production is maintained. Even when the voltage is reduced only by a small fraction by improvements in design, the saving may be considerable because of the scale of operation involved. Let us now look at each one of the components of cell voltage and see how they can be minimized to advantage.

The Nernst voltage is determined by the nature of the electrode process under discussion. If the raw material is given, there is nothing the engineer can do to minimize this part of voltage. The overvoltage due to electron transfer can be made smaller by using high concentrations of electroactive material and high electrode areas. The upper limit of concentration is the solubility of the reactant in whatever solvent is used. Another factor which may determine an upper limit for reactant concentration in molten salt cells is the melting point of the salt and its effect on the other components of the cell: if the temperature is raised, more energy is required for heating and corrosion becomes a greater problem. Therefore, sometimes the reactant is mixed with another salt, the mixture melts at a lower temperature and heating or corrosion problems are reduced at the expense of lowering the concentration of the reactant. The electrode area can be increased by using rough or porous electrodes. In porous electrodes, however, the mass transport within the pores may become a problem since there is a limit to the usable area of the electrode.

Concentration overvoltage can be minimized by using concentrated solutions and providing for suitable electrolyte agitation. Many forms of agitation have been used, including rotating electrodes and "windscreen wiper" devices. The latter also helps to remove gas bubbles from the electrode surface, reducing the loss of active area which always occurs when the product of the electrolysis is gaseous.

The resistance of the electrolyte can be minimized by placing the cathode and anode as close as possible. However, in cells with diaphragm, the distance between the electrodes cannot be very small. In other cells we must provide for the circulation of electrolyte between the electrodes and must avoid the situation where the two electrodes can be short circuited during operation. The resistance of the electrodes is minimized by careful choice of electrode materials and the smaller its specific resistance, the better. The electrodes themselves are massive blocks of metal or graphite, providing the shortest possible distance and largest area from the current source to the reaction site. This way of massive construction is also followed in the design of leads and bus-bars. The connections along the way, particularly those of bus-bars to electrodes, must be carefully designed and constructed because they can involve rather large non-uniform resistances. This involves not only loss of expensive power, but also local heating and erosion of the contact.

Introducing improvements in the design of cells may quite often result in more expensive cells. The minimization of concentration polarization by the use of extra agitators may serve as a good example. The engineers have a choice of either obtaining the current wanted by using higher voltage, or by providing suitable agitation of the electrolyte, which also consumes power. Clearly, if the voltage saved by agitation is greater than the power needed to provide it, the special agitators are a good investment. On the other hand, if the power saved is not very large, it is cheaper to use the higher cell voltage in maintaining the needed current. Very expensive, superbly designed cells are not practical for a production plant, again the balance between cost and efficiency must be maintained.

The performance of a cell is evaluated by the voltage needed for its operation and by its current efficiency. In the electrochemical industry it is expected that the desired reaction is the only one to take place at the suitable electrode and this expectation is often met in practice. However, it may happen that some mixing of the two reaction products or the two electrodes will take place, or that the product of one electrode reaction will diffuse to the surface of the other electrode. The result of these two possible processes is a loss in current efficiency in the same way as in the formation of side products. Thus, the current efficiency in the electrochemical industry is a measure of the efficiency of the diaphragm used in the cell.

Electrodes for electrochemical production processes should be good conductors, mechanically strong, free from chemical attack and should be efficient electrocatalysts for the reaction in question, i.e. this reaction should have maximum k_s value when taking place on that electrode. Most of the electrodes are solid metals or carbon (graphite); liquid electrodes of mercury, are used extensively as cathodes in brine electrolysis; molten lead electrodes are sometimes also used in molten-salt cells.

The shape of the electrode is indicated by its function and the nature of the product. In aluminium electrowinning, the cell body serves as the cathode; in the water electrolysis industry both of the electrodes are simple vertical plates. Naturally, the most difficult is the design of an electrode the product of which is gas. When bubbles are formed, they block a portion of the electrode area, decreasing efficiency and when they are dislodged, care must be taken to provide a free path for them to rise to the surface of the electrolyte. In some cases, notably the manufacture of fluorine, bubbles are not formed because the contact angle of fluorine-electrode and electrolyte is greater than 90°. The problems associated with this phenomenon will be discussed in the section on fluorine production. Naturally the most popular shape for gas producing electrode is the simple vertical plate. Sometimes, however, one wants to connect all of the plates to form a giant electrode in a giant cell. These connections are often perforated to allow the escape of gas.

Ideally, the current density on an electrode should be uniform, but this, in practice, is rarely the case. The edges of the electrodes carry a much larger current-density than the faces. In the laboratory this is of little consequence but in the industrial plant, this is the main reason for electrode, particularly anode, corrosion. When the electrode is made of a highly porous material, e.g. graphite, more electrolysis takes place in the pores at the edges. The pressure of the gaseous product formed breaks the graphite at that place, enlarging the holes. Increased electrolysis at the edges means also the increase of harmful side reactions, notably the oxidation of the graphite by electro-produced oxygen. Anode corrosion is, indeed, one of the main problems of the electrochemical industries, particularly the chlor-alkali and the aluminium ones.

The choice of cathode material is governed mainly by its corrosion resistance under conditions of operation and by the product of the cathode reaction. In aqueous solutions, if hydrogen is the cathodic product, one chooses metals with low hydrogen overpotential, such as iron or steel, bearing in mind the problem of hydrogen embrittlement. When the preferred reaction is not hydrogen evolution, mercury, lead or thalium are chosen because of their high overpotential toward hydrogen evolution. In electro-winning the metal itself serves as the cathode. Graphite and, less so, carbon are also used extensively as cathode materials. The choice of anode materials is much more restricted since most metals corrode rapidly under anodic conditions. The noble metals, notably platinum, is an excellent choice being both corrosion resistant and the best electrocatalyst. Platinum, however, is usually too expensive and in short supply, and platinum coated electrodes, notably of titanium, are often used instead. These electrodes have found considerable use in the chlor-alkali industry for the production of hypo-chlorites, chlorates and chlorine, in cathodic protection and electrowinning. Their increasing popularity is due to the availability of titanium in many forms and its strength, as well as to its corrosion resistant quality. The titanium anodes are always oxidized before coating with platinum, so that only the corrosion resistant oxide film is exposed in the cell. Such electrodes can be polarized to 100V in sulphuric acid at 20°C. The main problems of these electrodes is their initial cost and loss of platinum from the coating that may be quite expensive even though it is measured in micrograms per ton of product.

The only other metal which is used as an anode is lead and its alloy with 1% silver and 6-15% antimony. These are remarkably corrosion resistant in halide free solutions and are used mainly in the zinc and copper electro-winning industries.

Graphite and carbon have been used extensively as anodes in the electro-chemical industry for as long as this industry has existed. Graphite is usually preferred for chlorine chlorate, bromate and iodate production from their respective halide solutions. Graphite is also used as anode in electrowinning of sodium, lithium and magnesium. Non-graphitized carbon is not used as extensively as graphite, but is still applied in fluorine production and aluminium electrowinning. Graphite fulfills almost all requirements for a good electrode material, it is a good conductor, resistant to chemical attack, is fairly strong and is cheap. The problem of erosion of the electrodes during use is an acceptable penalty as their price is such that they can be periodically replaced.

The other two materials which have become useful as anode materials are compounds: lead dioxide and magnetite which are not oxidized any further under anodic conditions, indeed the stability of lead dioxide has been remarkable. Lead dioxide has been more widely used for anode construction than magnetite, its specific resistance is lower than many metals, it is hard and the oxygen overvoltage on it is the same as on platinum. Thus it is suitable as a substitute in many processes which initially required platinum anodes. The difficulty with both lead dioxide and magnetite anodes is their fabrication and this has delayed their development.

The use of diaphragms in commercial cells has already been mentioned. The ideal diaphragm should have the following properties: it should be permeable to ions but not to molecules; it should not add much to the cell resistance and at the same time should provide a complete barrier to the passage of gas bubbles and to diffusion; it should be homogeneous to provide even current density distribution; it should be chemically resistant to conditions in the cell; should have some mechanical strength and should be cheap. All these requirements clearly cannot be met in one diaphragm material and a compromise must usually be made.

The requirement of a barrier to the passage of molecules while not adding much to the cell resistance is clearly the one most difficult to realize. Low resistance requires a large area of pores in the diaphragm, but being a barrier requires that the pores should be very small. When the solutions in the anode and cathode compartments are very different, as in the electrolysis of brine, it helps to use a filtering diaphragm. Here the level of the liquid in the anode compartment is kept higher than that in the cathode compartment. Thus, by forcing diffusion from the anode to cathode compartments, the highly caustic solution is prevented from reaching the anode.

Diaphragms are usually not expected to last for very long periods because of the inevitable mixing that takes place in the diaphragm which may result in the formation of precipitates in the pores, blocking any passage of electrolyte.

The commonest materials for the production of diaphragms are asbestos, ceramic, plastic materials, such as P.V.C., polypropylene and Terylene, graphite cloths and stainless steel. The two latter materials are conductors and, therefore, acquire a potential respective to the solution when placed in the cell. They may, if not properly designed and placed, function as extra electrodes in the cells. This should always be borne in mind when using conducting diaphragms.

The electrolytes used in industrial processes should possess good electrical conductivity, should dissolve the reactant in high concentrations and should cause as few corrosion problems as possible. If the product is also soluble in the electrolyte, the process is smoother, but there is the additional problem of separating the product. The electrolytes used are either aqueous solutions or molten salts; no organic solvents have been used commercially.

Cell bodies are often made from cast iron or mild steel with or without rubber lining. Resin bonded glass fibre and plastic materials may replace some of the older materials for cell construction. Plastic gaskets are certainly used instead of tars and pitches for sealing.

We have very briefly surveyed some of the problems of the application of electrochemical processes to the industrial plant. We have listed some of the general features of construction found useful by experiment and experience. In the following section we shall deal with some of the more important processes of the electrochemical industry.

2. Examples of Electrochemical Processes

(a) The chlor-alkali industry is, together with aluminium production, the most important electrochemical industry. Its importance lies in the very wide demand for chlorine and its oxy compounds in the world market. The only way to oxidize chloride ions efficiently is by the use of electric current. The chlor-alkali industry includes the production of caustic soda, metallic sodium, gaseous chlorine, hypochlorite, chlorate and perchlorate, bromate and iodate. All these plants are often found on the same site. The raw material is mostly aqueous brine of high concentration, the solution of NaCl is almost saturated.

Two types of cells are used for the electrolysis of brine: the mercury cell and the diaphragm cell. In the mercury cell the cathode is mercury, the anode is graphite and the cell reaction is

$$2NaCl = 2Na(Hg) + Cl_2 \qquad\qquad \text{XL}$$

The amalgam formed (0.3% Na usually) is decomposed in "denuders"

$$2Na(Hg) + 2H_2O = H_2 + 2NaOH \qquad\qquad \text{XLI}$$

which use graphite blocks or balls as catalysts. The caustic soda obtained is very pure. Diaphragm cells have mild steel cathodes and graphite anodes and the reaction is

$$2NaCl + 2H_2O = 2NaOH + H_2 + Cl_2$$

The caustic produced also contains up to 50% NaCl. The diaphragm cells are much cheaper to install than the mercury cell, the price of mercury being a determining factor. The energy required to form the amalgam is also higher than that needed for the formation of hydrogen on steel, increasing the power consumption of Hg cells. The advantage or disadvantage of the use of any one of the two cells is thus determined by the market for pure caustic soda.

The production of chlorates from brine is now increasing due to its use in the paper and pulp industry. The overall reaction

$$NaCl + 3H_2O = NaClO_3 + 3H_2 \qquad\qquad \text{XLII}$$

is carried out in cells with cathodes of mild or stainless steel and anodes of either platinum coated titanium, lead dioxide, graphite or magnetite. The cells have no diaphragm and the products are circulated so that the chlorate is formed outside the cell via the following reactions

$$Cl_2 + H_2O = HClO + H^+ + Cl^- \qquad\qquad \text{XLIII}$$

$$Cl_2 + 2OH^- = ClO^- + H_2O + Cl^- \qquad\qquad \text{XLIV}$$

which are fast. The slow step is

$$2HClO + ClO^- = ClO_3^- + 2H^+ + 2Cl^- \qquad\qquad \text{XLV}$$

This reaction is favoured by low temperatures and mildly acid conditions. After separation of chlorate the resulting chloride solution is recirculated.

Perchlorates are manufactured by a similar process.

Hypochlorites are made by electrolysis in small plants where sea water is used as brine. On a large scale, it is made by dissolving chlorine in caustic soda in a plant adjacent to, but separate from, the chlorine plant.

All sodium manufactured is obtained by the electrolysis of sodium chloride melt. The electrolyte, however, is not pure sodium chloride but a mixture of sodium, calcium and barium chlorides. The cell is constructed of ceramic lined iron and contains four graphite anodes and two iron cathodes. Both sodium and chlorine rise to the surface, hence the need for a diaphragm, usually of iron. Corrosion of the graphite anodes and of the diaphragm is a persistent problem of operation of these cells. Another problem is the calcium which is formed with the sodium and sometimes clogs the outlet channels of the cell.

(b) Fluorine manufacture. Fluorine is by far the strongest oxidizing agent known; clearly it cannot be produced from its compounds using chemical processes. The only way to obtain elemental fluorine is by electrochemical oxidation of its compounds. There are many problems involved in the manufacture of fluorine, most of which originate in this very reactive nature of the product. The overcoming of these problems makes very interesting reading but is outside the scope of this section.

The electrolyte used in fluorine cells is a molten mixture of the approximate composition $KF \cdot 2HF$, at 80–100°C. The anodes are carbon and the cathodes are usually steel. The fluorine liberated at the anode is expected to leave as gas, but it does not form bubbles on the electrode in the usual way, giving rise to what is known in the fluorine industry as "polarization".

When a fluorine cell is operated at fair current densities for some short time, the current drops to a small fraction of its former value and stays there. If the potential is increased very much, current starts to flow again accompanied by intense heating and combustion of the superficial layers of the anode. This effect is known as the "anode effect".

In most gas forming electrochemical processes, the electrolyte wets the electrode, i.e. the contact angle between the electrode and electrolyte at the gas surface is very much smaller than 90°. Molten fluoride mixture also wets carbon in a similar way but when the carbon is made anodic, an intercalation carbon fluorine compound is formed at the surface. This compound is not wetted by the fluoride melt, the contact angle of the melt, carbon fluorine compound and fluorine is approximately 150°. Thus, any bubbles formed at the electrode surface will not be round but lenticular in shape and when they grow do not break from the surface, but slide upwards and join to other bubbles to form large gas covered areas. These, in turn, cover a large portion of the electrode surface and the result is a drop in cell current. When the current drops, bubbles cease to form and whatever gas there is on the electrode is absorbed into the electrode pores, freeing the electrode from the gas film. An equilibrium is reached at some low current density, this is "polarization".

There are two ways by which "polarization" can be overcome. One is to use carbon electrodes of rather large pores where the gas formed leaves the cell via the interconnecting pores and not via bubbles. This way depends on the diffusion of the gas through pores and is, therefore, too slow to employ industrially. The other method is to add "wetting agents" to the cell, a most important one being a suspension of high-valence nickel compounds. These compounds, which form in water-free cells from the corrosion of the diaphragms (which are made from nickel containing alloy) reduce the contact angle to a value very much smaller than 90°, thus promoting the formation of normal, spherical bubbles.

While raising very interesting problems, elemental fluorine is currently used on a large scale only in the atomic energy industry to make UF_6. The latter is fed to diffusion separation plants the product of which is U^{235}-rich metal for use in nuclear reactors. Some other uses, such as a powerful oxidant in rocketry or for the manufacture of fluorine rich compounds are envisaged.

Highly fluorinated organic compounds can sometimes be obtained by electrolysis of the organic compound solution in HF or HF and KF mixtures, using nickel anodes and steel cathodes. The scope of the process is very wide, its main attractive feature being that many functional groups, particularly the carboxylic acid group and the ether group, are not destroyed. Many other functional groups, such as alcohols or amines, are oxidized. The yield of any particular compound has been quite poor and the process is now the subject of academic and industrial research.

(c) Electrowinning of metals. All metals, when recovered from their ores, must be reduced either by carbon, hydrogen or other reducing agents, or by electrolysis. The decision as to which process to use is based on the required purity of the metal produced and on economic considerations. The advantages of electrowinning are (a) the ability to use poorer ores more efficiently and (b) the reducing power of cathodes, which can produce even the most electropositive metals from their compounds. Electrowinning processes are divided to those using aqueous electrolytes and those using molten salt electrolytes. The former group includes those metals which are less electropositive than manganese; the molten salt group includes the alkainei, alkaline earth, the earths (B, Al. etc.), actinides and Ti, Nb, Ta of the transition metals.

The general process of aqueous electrowinning includes four steps: (a) Conversion of the ore to an acid soluble form, such as oxide, by roasting, (b) leaching of the ore by acid and purification of the solution, (c) electrolysis and (d) recycle of the acid liberated at the anode compartment. If the ore is already in an oxide form the ore preparation step (a) is eliminated. Step (b) is the most crucial in the process. The solution must be free from metals more noble than the one to be electrowon or these noble metals will deposit preferentially. The solution must often be free from more electropositive metals since the latter may tend to form undesirable precipitates in the cell. This purification step has the result that the final product is very pure, usually much more so than in chemical reduction processes. A common impurity, however, is hydrogen.

Electrowinning in molten salts is the only way to obtain some metals in the elementary state. The process, however is not straightforward and involves many problems. Corrosion of the cell components and electrodes is a persistent very serious problem. If the metal deposits as a solid, separation may be difficult; growth of dendrites may cause short-circuit between the anode and cathode. Solubility of the metal in the melt may be a source of current inefficiency.

Of the numerous electrowinning processes, those for zinc and aluminium will be briefly described.

Electrowinning of zinc is from sulphuric acid leach. The solution is purified from more noble metals by two methods: one is to add scrap iron, oxidize it to the ferric state which is precipitated, carrying down many impurities. The other method is to add zinc dust, thus precipitating the more noble metals. Cadmium, as well as other metals, are usually recovered at this stage. The electrolysis is between lead-silver anodes and aluminium cathodes; the product is generally 99·9% pure.

Aluminium is by far the most important metal produced by electrowinning and its importance in many fields is undisputed. Since all aluminium is produced by electrowinning, the importance of this process in this context is clear.

Aluminium ore is chiefly bauxite. This is converted to dry alumina by the Bayer process and the alumina, which is usually of purity of 99%, is dissolved in molten cryolite in the electrolysis cell. The cell is made of steel lined with insulating material which, in turn, is lined with carbon. This carbon supports the liquid aluminium cathode. The vertical anode is also made of carbon and the cell is kept at 950°C by resistive heating. The operation of the cell is by batches. the solution which contains 5% alumina is depleted to about 1%, when the voltage across the cell rises abruptly. When this happens, more alumina is added to the cell. The molten aluminium is tapped periodically. The most serious problems of this process are connected with rapid erosion of the carbon electrodes—0·5 kg carbon per kg of aluminium. Thus the position of the anode has to be continuously adjusted to keep the distance between the electrodes constant. The carbon lining of the cell also suffers attack; it is distorted by the thermal gradient between its surface and the steel shell and by the penetration of sodium ions from the melt into the lining. These distortions cause heaving of the lining and finally cracking. When the carbon lining is cracked, the molten aluminium attacks the current collector bars, the insulated lining and also the steel shell. It is this failure of the carbon lining that limits the life of a cell to three–five years.

(*d*) Electrorefining of metals. This is the process where the impure metal anode corrodes and the metal is deposited on the cathode in a pure form. The relationship of this branch of technology to electro-deposition and electrowinning is apparent. The principle of electrowinning is that the anode gives off as the only impurities metals more electropositive to the one to be refined; the nobler metals do not dissolve. The base metals, however, are not deposited on the cathode because the potential of the latter (compared to the s.h.e.) is too positive. Thus the impurities stay either on the anode or in the electrolyte. This latter fact creates the need for rigorous purification of the electrolyte and, hence, its efficient circulation.

The reversible potential of the electrorefining cell is, clearly, zero. Overpotentials of various kinds being, as always, present, the actual operating potential is not zero but is still very low, so that power consumption is not the major expense in running this refinery. When metals are needed in a very pure form or when separation of other valuable metals is wanted, electrorefining is unquestionably the process to use.

A simple electrolytic refinery consists of the following parts: (*a*) the cells, (*b*) an electrolyte storage and circulation system which may also include a purification system, (*c*) equipment for slime recovery and (*d*) cranes for handling the electrodes. The cell itself is usually a rectangular tank made of suitably resistant material and the electrolyte usually contains several addition agents to promote anodic dissolution, conductivity, constant acidity or smooth adherent deposits. These are the same kinds of addition agents used in electrodeposition. It is very important to obtain pure cathodes that are free of scale, lumps and nodules because often the cathodes are sold directly.

F. Summary

This last chapter attempted to present the more important or more promising applications of electrodics in technology. Clearly, of the dozen or so volumes used in the preparation of this chapter, much pertinent and interesting information was omitted, often quite arbitrarily. The interested reader is referred to the Bibliography. It may be found that in very many cases, the most crucial data for the understanding of the process may be missing. This is sometimes due to the different viewpoints of basic scientists and engineers—what seems of great importance to one group may seem trivial to another. Sometimes, however, important data are kept secret by the producing companies, for obvious reasons.

Bibliography

R. N. Adams, *Electrochemistry at Solid Electrodes*, Marcel Dekker Inc., New York, 1969.

J. O'M. Bockris and A. K. N. Reddy, *Modern Electrochemistry*, Vol. 2, Plenum Press, New York, 1970.

B. E. Conway, *Theory and Principles of Electrode Processes*, The Ronald Press Co., New York, 1965.

P. Delahay, *New Instrumental Methods in Electrochemistry*, Interscience Publishers, New York, 1954.

P. Delahay, *Double Layer and Electrode Kinetics*, Interscience Publishers, New York, 1965.

B. B. Damaskin, *The Principles of Current Methods for the Study of Electrochemical Reactions*, McGraw–Hill, 1967.

B. B. Damaskin, O. A. Petrii and V. V. Batrakov, *Adsorption of Organic Compounds on Electrodes*, Plenum Press, New York, 1971.

K. Denbigh, *The Principles of Chemical Equilibrium*, Cambridge at the University Press, 1961.

T. Erdey-Gruz, *Kinetics of Electrode Processes*, Adam Hilger Ltd., London, 1972.

U. R. Evans, *An Introduction to Metallic Corrosion*, 2nd Edn., Edward Arnold, 1963.

G. Eisenman, " The Electrochemistry of Cation Sensitive Glass Electrodes ", in *The Glass Electrode*, Interscience Publishers, New York.

J. Heyrovsky and J. Kuta, *Principles of Polarography*, Academic Press, New York, 1966.

J. Koryta, J. Dvorak and V. Bohackova, *Electrochemistry*, Methuen & Co. Ltd., London, 1970.

A. T. Kuhn, ed., *Industrial Electrochemical Processes*, Elsevier, Amsterdam, 1971.

L. L. Leveson, *Introduction to Electroanalysis*, Butterworths, London, 1964.

C. K. Mann and K. K. Barnes, *Electrochemical Reactions in Non-Aqueous Systems*, Marcel Dekker Inc., New York, 1970.

L. Meites, *Polarographic Techniques*, 2nd edn., Interscience Publishers, New York, 1965.

C. H. Page and P. Vigoureux, ed., *The International System of Units (SI)*, NBS Special Publication 330, 1972 Edition.

G. R. Pallin, *Electrochemistry for Technologists*, Pergamon Press, Oxford, 1969.

R. A. Robinson and R. H. Stokes, *Electrolyte Solutions*, Butterworths Scientific Publications, London, 1959.

K. J. Vetter, *Electrochemical Kinetics, Theoretical Aspects*, Academic Press, New York, 1967.

J. M. West, *Electrodeposition and Corrosion Processes*, 2nd edn., Van Nostrand, 1972.

J. R. Wiliams, *An Introduction to Fuel-Cells*, Elsevier, Amsterdam, 1966.

Subject Index

Numbers in italics show the location of a definition or of a section on the subject. Certain key words such as anode, cathode, electrode etc. appear very frequently. The Index shows only the location of their definition.